CONTENTS

ISBN: 9780170389426

Glossary

Make your own glossary of key terms:

Term	Definition	Picture/Example
Variable		
Coefficient		
Constant		
Dependent		
Consistent		
Inconsistent		
Plane		
Parallel		

What are simultaneous equations?

- Simultaneous equations are sets of equations with **two or more** unknowns or **variables**.
- 'Simultaneous' means '**at the same time**'.
- You need to be able to solve **two** or **three** equations that are true 'at the same time'.
- In order for the equations to be solved, the number of equations must be at least equal to the number of variables: 2 variables $\Rightarrow$ 2 equations; 3 variables $\Rightarrow$ 3 equations, etc.
- The sets of equations may have **no** solutions, **one** solution or **many** solutions.

ISBN: 9780170389426

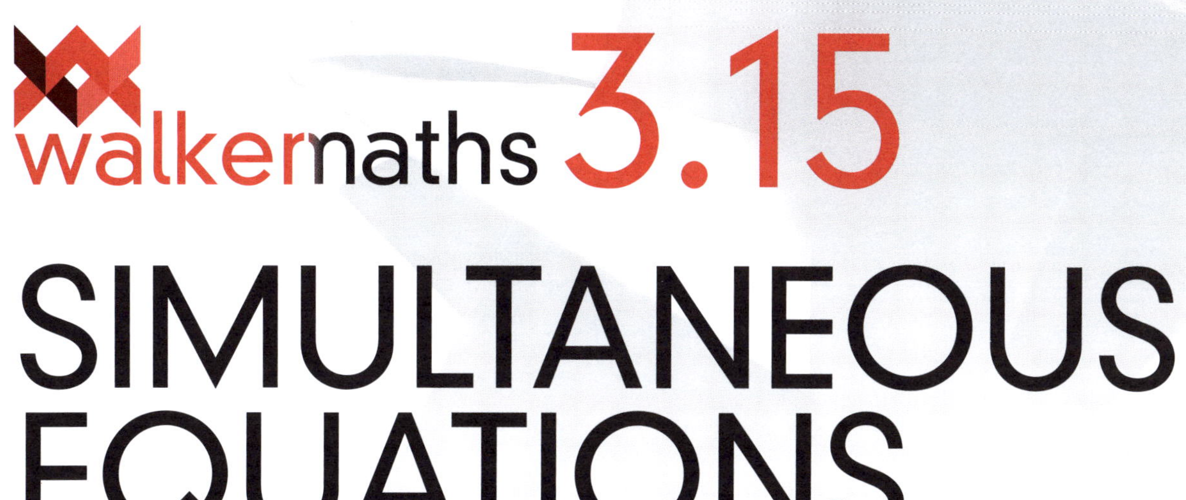

SIMULTANEOUS EQUATIONS

NCEA Level 3 Internal

Charlotte Walker and Victoria Walker

Walker Maths 3.15 Simultaneous Equations
1st Edition
Charlotte Walker
Victoria Walker

Editor: Eva Chan
Cover and text design: Cheryl Smith, Macarn Design
Production controller: Siew Han Ong and Jennifer Foo

Acknowledgements
Cover photo courtesy of Shutterstock.

For product information and technology assistance,
in Australia call **1300 790 853**;
in New Zealand call **0800 449 725**

For permission to use material from this text or product, please email
aust.permissions@cengage.com

National Library of New Zealand Cataloguing-in-Publication Data
A catalogue record for this book is available from the National Library of New Zealand.

978 0 17 038942 6

Cengage Learning Australia
Level 7, 80 Dorcas Street
South Melbourne, Victoria Australia 3205

For learning solutions, visit **cengage.co.nz**

Printed in China by 1010 Printing International Limited.
12 25

Solving 2 x 2 simultaneous equations

- 2 x 2 simultaneous equations have **two variables** and there must be at least **two equations**.
- An equation with **two** variables represents a **line**.
- By solving these equations we are attempting to find their **point of intersection**.
- There are three ways of solving simultaneous equations: by plotting, by substitution or by elimination.

1 Plotting

Example: Plot the graphs of $y = x + 1$ and $2y + 2x = 10$.

You may find it easier to rearrange this so that $y = \ldots\ldots$

$$2y = 10 - 2x$$
$$\therefore y = 5 - x$$

Step 1: Create tables for values of x and y.

$y = x + 1$

x	x + 1	y	Point
0	0 + 1	1	**(0, 1)**
1	1 + 1	2	**(1, 2)**
2	2 + 1	3	**(2, 3)**
3	3 + 1	4	**(3, 4)**
4	4 + 1	5	**(4, 5)**

$y = 5 - x$

x	5 – x	y	Point
0	5 – 0	5	**(0, 5)**
1	5 – 1	4	**(1, 4)**
2	5 – 2	3	**(2, 3)**
3	5 – 3	2	**(3, 2)**
4	5 – 4	1	**(4, 1)**

Step 2: Plot the points.

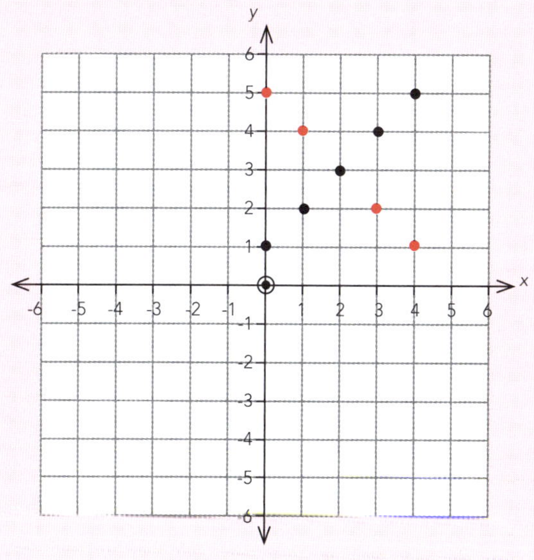

Step 3: Rule the lines.

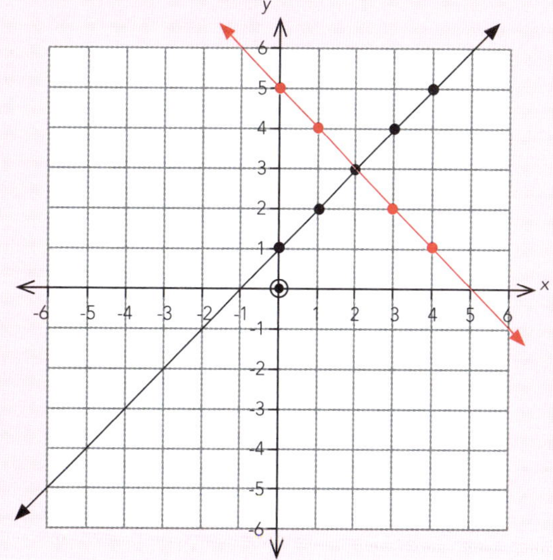

Step 4: Read the intercept off the graph. **The intercept is (2, 3).**

ISBN: 9780170389426

Plot the following pairs of lines and read the point of intersection.

1 $y = 2x + 1$

____

____

____

x		y	Point
0			
1			
2			
3			
4			

$2x + y = 7$

____

____

____

x		y	Point
0			
1			
2			
3			
4			

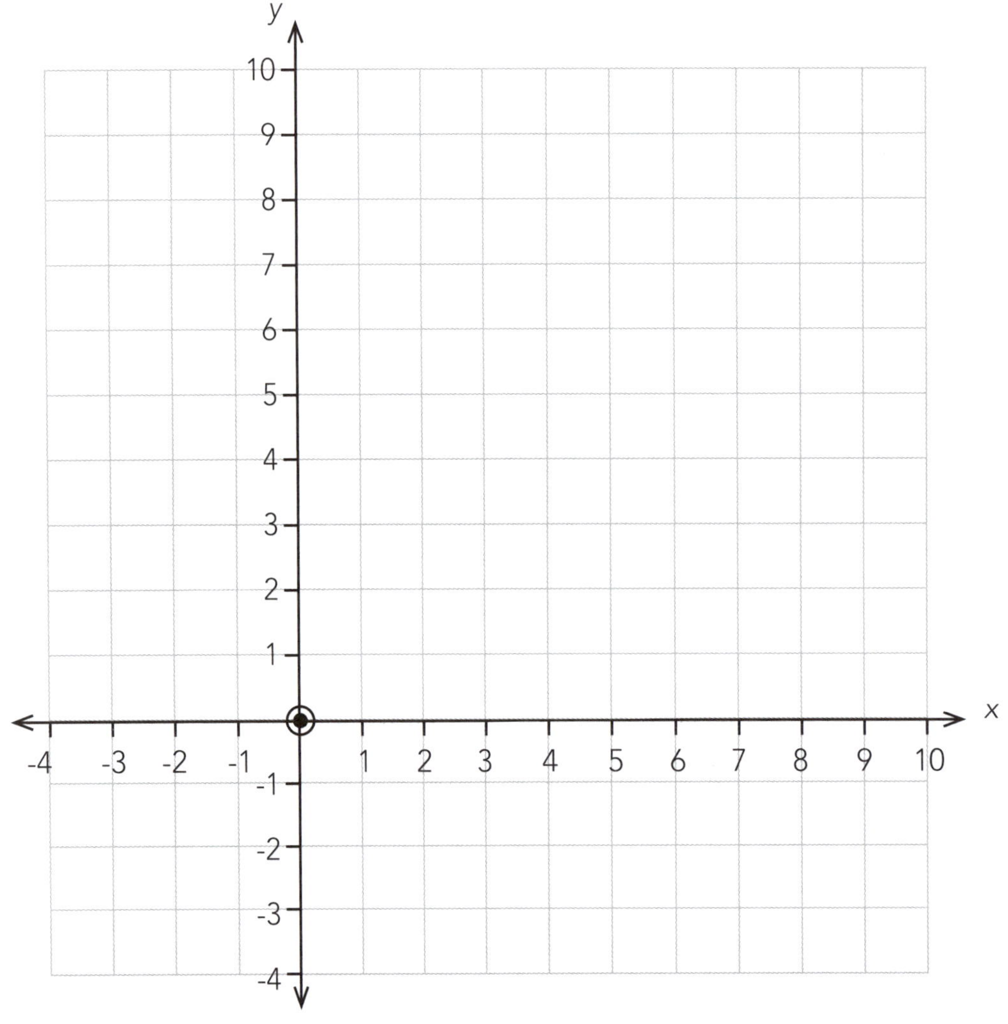

The intercept is (,   ).

 ISBN: 9780170389426

2 $y = 3x - 1$

$x = 2y - 8$

x		y	Point
0			
1			
2			
3			
4			

x		y	Point
0			
1			
2			
3			
4			

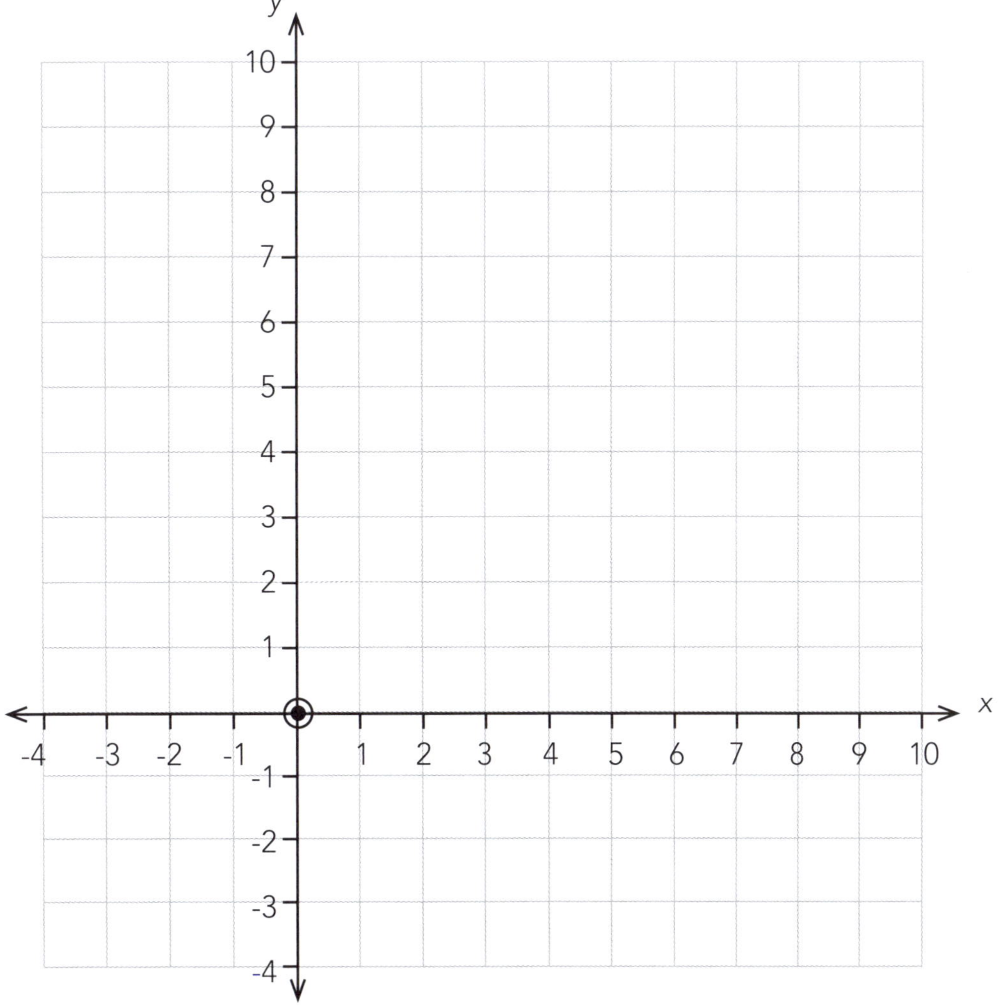

The intercept is (,   ).

2 Calculation

- There are **two methods** for solving by calculation.
- Which one you use depends on the structure of the equations.
- **Both** methods **will always work**, but selecting the best method will make your calculations easier.
- You should **number** each equation, and **say what you are doing** at each step.

a Substitution

- Substitution is easiest where one of the equations is expressed as

x = ... or **y = ...**

- As the name suggests, you **substitute** the **x = ...** or **y = ...** into the **other** equation.

Examples:

1 Solve the equations $y = 3x - 3$ and $2x + y = 7$.

$$y = 3x - 3 \quad ①$$
$$2x + y = 7 \quad ②$$

Number the equations.

Substitute ① into ②:

$$2x + (3x - 3) = 7$$
$$2x + 3x - 3 = 7$$
$$5x - 3 = 7$$
$$5x = 10$$
$$x = 2$$

From equation ① we know that **y** and **3x – 3** are equal.

Substitute for y in ①:

$$y = 3(2) - 3$$
$$y = 3$$

The intercept is **(2, 3)**.

2 Solve the equations $x = 2y - 1$ and $3x - y = 12$.

$$3x - y = 12 \quad ①$$
$$x = 2y - 1 \quad ②$$

Substitute ② into ①:

Select ② because it has the form x = ...

$$3(2y - 1) - y = 12$$
$$6y - 3 - y = 12$$
$$5y - 3 = 12$$
$$5y = 15$$
$$y = 3$$

Substitute for y in ②:

$$x = 2(3) - 1$$
$$x = 5$$

Select whichever equation is easier for substitution.

The intercept is **(5, 3)**.

ISBN: 9780170389426

Solve the following simultaneous equations using substitution.

1
$$x = y + 1$$
$$x + 3y = 21$$

2
$$y = 2x + 1$$
$$x + 2y = 27$$

3
$$x = 3y + 9$$
$$6y - x = 3$$

4
$$y = 4x - 2$$
$$9x - 2y = 8$$

5
$$3g + 2h + 6 = 0$$
$$h = 11 + 2g$$

6
$$2x + y - 1 = 0$$
$$x = 5 - y$$

7
$$x + y = 1$$
$$2x + 3y = 4$$

8
$$5y - x = y + 1$$
$$y = x + 7$$

9
$$4a - 8b = 12$$
$$-3a + b = 1$$

10
$$2x = 8y + 4$$
$$3x - 2y = -17$$

ISBN: 9780170389426

b Elimination

- Elimination is easiest when the two equations have the **same structure**.

For example:

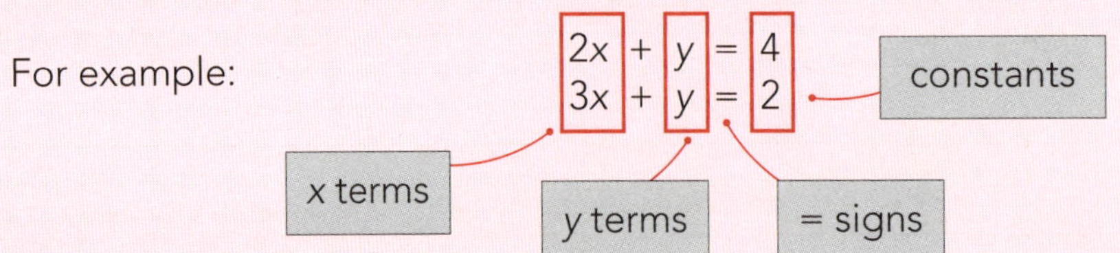

- You solve these by multiplying one or both equations by a constant in order to make either the x terms or the y terms the **same size**.
- Then **add or subtract** the two equations to produce an equation with **one variable** only.
- You should **number** each equation, and **say what you are doing** at each step.

Examples:

1 Solve the equations $2x + y = 4$ and $3x + y = 2$.

	$2x + y = 4$	①
	$3x + y = 2$	②
Subtract ① and ②:	$-x = 2$	
	$\mathbf{x = -2}$	
Substitute for x in ①:	$2(-2) + y = 4$	
	$\therefore\ \mathbf{y = 8}$	

The intercept is **(-2, 8)**

2 Solve the equations $8x + 3y = -5$ and $5x + y = 3$.

	$8x + 3y = -5$	①
	$5x + y = 3$	②
Multiply ② by -3:	$-15x - 3y = -9$	③
Add ① and ③:	$-7x = -14$	
	$\therefore\ \mathbf{x = 2}$	
Substitute for x in ②:	$5(2) + y = 3$	
	$\therefore\ \mathbf{y = -7}$	

The intercept is **(2, -7)**

3 Solve the equations $5x - 2y = 10$ and $7x - 3y = 13$.

	$5x - 2y = 10$	①
	$7x - 3y = 13$	②
Multiply ① by -3:	$-15x + 6y = -30$	③
Multiply ② by 2:	$14x - 6y = 26$	④
Add ③ and ④:	$-1x = -4$	
	$\therefore\ \mathbf{x = 4}$	
Substitute for x in ①:	$5(4) - 2y = 10$	
	$-2y = 10 - 20$	
	$\therefore\ \mathbf{y = 5}$	

You need to multiply both equations by constants.

The intercept is **(4, 5)**

ISBN: 9780170389426

Solve the following simultaneous equations.

1

$4x - y = 6$
$-3x + y = -2$

2

$-5x + 3y = 23$
$2x + 3y = 16$

3

$2x - 5y = 10$
$2x + 3y = 6$

4

$3x + 2y = 15$
$-12x + 10y = 75$

5

$2x + 3y = 6$
$4x + 5y = 4$

6

$-6y + 3x = 3$
$5x + 2y = 14$

7

$6x + 5y - 16 = 0$
$4x + 3y = 10$

8

$-3x = -36 + 2y$
$5x + 10y = 32$

9

$1.5a + b = 1.5$
$7a - 3b = -39$

10

$-12 = 3s + 6t$
$6s + 13t = 25$

ISBN: 9780170389426

Mixing it up

Use either substitution or elimination to solve these.

1
$$x = 4 - 9y$$
$$2x + 3y = 53$$

2
$$3x + 3y = 9$$
$$-6x - 3y = 6$$

3
$$4x - 3y = 15$$
$$-2x + 9y = 21$$

4
$$2x + 6y = 0$$
$$4x - 3y - 12 = 0$$

5
$$x - 2y = -6$$
$$2x + y = 11$$

6
$$3y + 9x = 36$$
$$8x + 6y = 17$$

7
$$-x + 2y = 3$$
$$x = 5 - 3y$$

8
$$3x = 6 + 2y$$
$$7x - 3y = -6$$

9
$$6x + 6y = 12$$
$$5x + y = 2$$

10
$$-2 = 4x + 1.5y$$
$$5x + 4y = 6$$

ISBN: 9780170389426

Solving 3 x 3 simultaneous equations

- 3 x 3 simultaneous equations have **three variables** and there must be at least **three equations**.
- An equation with **two** variables represents a **line**.
- An equation with **three** variables represents a **plane**.
- In a 3 x 3 system of simultaneous equations, the **planes** usually **meet** (intersect) at **one point**.
- By solving these equations we are attempting to calculate their **point of intersection**.
- They can be solved only by **substitution** or **elimination**.

1 Substitution

If one of the equations is written as **x = ...** or **y = ...** or **z = ...**, it is probably easier to solve using **substitution**.

Example:

Solve the equations

$x = -5 - 2y - 3z$ ①

$3x + y - 3z = 4$ ②

$-3x + 4y + 7z = -7$ ③

Substitute ① into ②:

$3(-5 - 2y - 3z) + y - 3z = 4$

$-15 - 6y - 9z + y - 3z = 4$

$-5y - 12z = 19$ ④

Substitute ① into ③:

$-3(-5 - 2y - 3z) + 4y + 7z = -7$

$15 + 6y + 9z + 4y + 7z = -7$

$10y + 16z = -22$ ⑤

Solve ④ and ⑤:

④ x 2 $\quad -10y - 24z = 38$ ⑥

⑤ $\quad 10y + 16z = -22$ ⑤

⑤ + ⑥ $\quad -8z = 16$

$\mathbf{z = -2}$

Substitute $z = -2$ into ④:

$-5y - 12(-2) = 19$

$-5y = -5$

$\mathbf{y = 1}$

Substitute $z = -2$ and $y = 1$ into ①:

$x = -5 - 2(1) - 3(-2)$

$\mathbf{x = -1}$

The order in which you find the answers does not matter.

So the point of intersection of these three planes is **(-1, 1, -2)** or where **x = -1, y = 1, z = -2**.

ISBN: 9780170389426

Use substitution to solve these equations to find the point of intersection.

1

$$x = 1 - 3y + 2z$$
$$2x - y - 3z = 13$$
$$3x + 2y - z = 2$$

2

$$y = 4 - 2x + z$$
$$4x + 2y + 6z = 8$$
$$-7x - 3y - 9z = -12$$

3

$$x = y + z$$
$$-x + 2y - z = 14$$
$$-6x + 2y - z = -11$$

4

$$x - y - z = 16$$
$$-2x + 2y + 3z = -18$$
$$3x + 3y - 3z = 24$$

 ISBN: 9780170389426

2 Elimination

Once again, **elimination** is easiest when the two equations have the **same structure**.

For example:

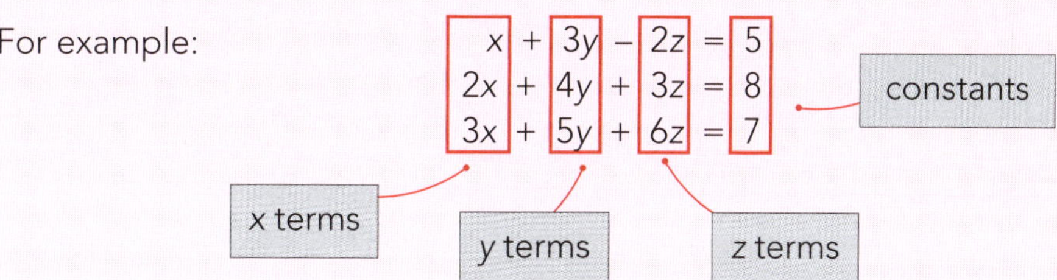

Example:

Solve the equations

$$x + 3y - 2z = 5 \quad ①$$
$$2x + 4y + 3z = 8 \quad ②$$
$$3x + 5y + 6z = 7 \quad ③$$

Eliminate the **same** variable from two **different** pairings of the equations:

From ① and ②:

① x 2 $\quad 2x + 6y - 4z = 10 \quad ④$

② $\quad 2x + 4y + 3z = 8 \quad ②$

④ − ② $\quad \mathbf{2y - 7z = 2} \quad ⑤$

From ① and ③:

① x 3 $\quad 3x + 9y - 6z = 15 \quad ⑥$

③ $\quad 3x + 5y + 6z = 7 \quad ③$

⑥ − ③ $\quad \mathbf{4y - 12z = 8} \quad ⑦$

From ⑤ and ⑦:

⑤ $\quad 2y - 7z = 2 \quad ⑤$

⑦ ÷ -2 $\quad -2y + 6z = -4 \quad ⑧$

⑤ + ⑧ $\quad \mathbf{z = 2}$

Substitute $z = 2$ into ⑤:

$$2y - 7(2) = 2$$
$$2y = 16$$
$$\mathbf{y = 8}$$

Substitute $z = 2$ and $y = 8$ into ①:

$$x + 3(8) - 2(2) = 5$$
$$x + 24 - 4 = 5$$
$$\mathbf{x = -15}$$

So the point of intersection of these three planes is at **(-15, 8, 2)**, or where **$x = -15$, $y = 8$, $z = 2$**.

ISBN: 9780170389426

Use elimination to solve these equations to find the point of intersection.

1

$$x - 2y - z = 4$$
$$2x + 2y + 3z = -2$$
$$-3x + 2y + 3z = 8$$

2

$$x - 4y + 2z = 12$$
$$2x + y + 3z = 15$$
$$3x + 3y - 5z = 21$$

3

$$3x + 2y + z = 0$$
$$5x + 6y + 7z = 8$$
$$-2x + 3y + 4z = 12$$

4

$$x + y + z = 30$$
$$2x - y + 3z = 19$$
$$3x - 2y + z = 7$$

ISBN: 9780170389426

Mixing it up

Use either elimination, substitution or a mixture to solve these equations.

1

$$x - 2y + 2z = -10$$
$$x - 4y - 5z = 6$$
$$-x + 4y + 2z = 3$$

2

$$x + y + z = 1$$
$$-2x + 4y + 8z = 10$$
$$9x + 5y + 2z = -16$$

3

$$-x - 6y + 4z = 3$$
$$y + 5z = 16$$
$$2x - 3y + 2z = 9$$

4

$$-5x + 3y + 4z = 0$$
$$x - 3y - 3z = 25$$
$$4x + y - z = 15$$

ISBN: 9780170389426

5

$$-x + y + 3z = -14$$
$$2x + 5y - 2z = 3$$
$$x + 4y = 2$$

6

$$7x + 4y + 2z = 2$$
$$-2x + 3y + 3z = -4$$
$$3x + 5y + 6z = 12$$

7

$$2x - 3y + 0.5z = 10$$
$$x + 5y - 4z = 4$$
$$-3x - 2y + 6z = 8$$

8

$$2y + 3z = 4$$
$$5x - 3y + 2z = 0$$
$$-4x + 4y + z = 6$$

ISBN: 9780170389426

Using your graphics calculator

1 Solving 2 x 2 simultaneous equations

⟶ **Menu**

⟶ **Equa**

⟶ **F1: Simultaneous**

⟶ **F1: 2 Unknowns**

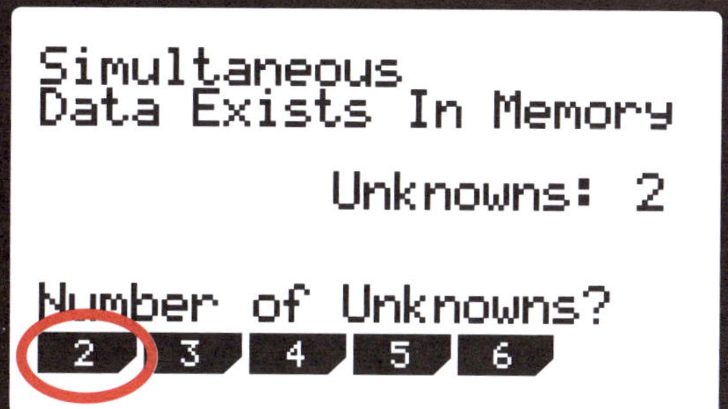

This shows you the format in which your equation must be.

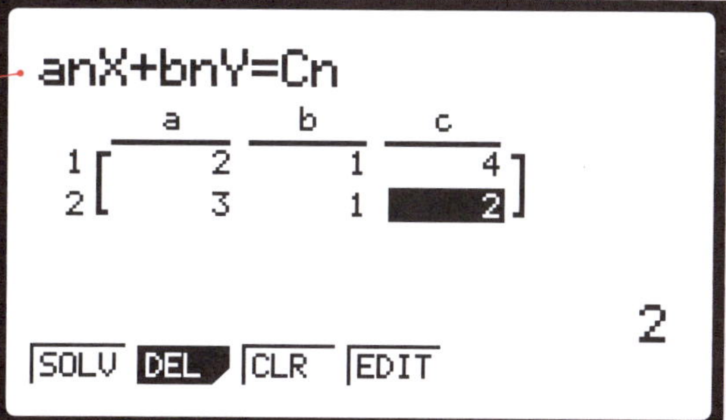

Example

$2x + y = 4$ ①

$3x + y = 2$ ②

⟶ **F1: Solve**

anX+bnY=Cn

X −2

Y 8

−2

REPT

So the point of intersection of the lines is (-2, 8).

Try this one:

$y = 3x - 3$ ①

$2x + y = 7$ ②

This is a little trickier as you will need to rearrange before entering it into your calculator.

____

____

____

You should get a solution (2, 3).

ISBN: 9780170389426

Use your calculator to solve the following simultaneous equations.

1
$$25x + 10y = 0$$
$$-x + 6y = 32$$

2
$$3x = 9 + 4y$$
$$-2x + 3y = 10$$

3
$$x + y = 6$$
$$5 = -x - 2y$$

4
$$5x + 5y = 15$$
$$-3x = 3 + 4y$$

5
$$2 = 12x + 3.5y$$
$$-6x + 2y = 5$$

6
$$6y + 3x = -3$$
$$2x - 3y = -2$$

7
$$6x - 2y + 9 = 0$$
$$x + 6y - 8 = 0$$

8
$$3x = 4 + 2y$$
$$-5y = 13 - 6x$$

9
$$6x = -5y + 5$$
$$3x + 5y = 2$$

10
$$-6 = -3y + 6x$$
$$5x + y + 5 = 0$$

ISBN: 9780170389426

2 Solving 3 x 3 simultaneous equations

→ **Menu**

→ **Equa**

→ **F1: Simultaneous**

→ **F2: 3 Unknowns**

This shows you the format in which your equation must be.

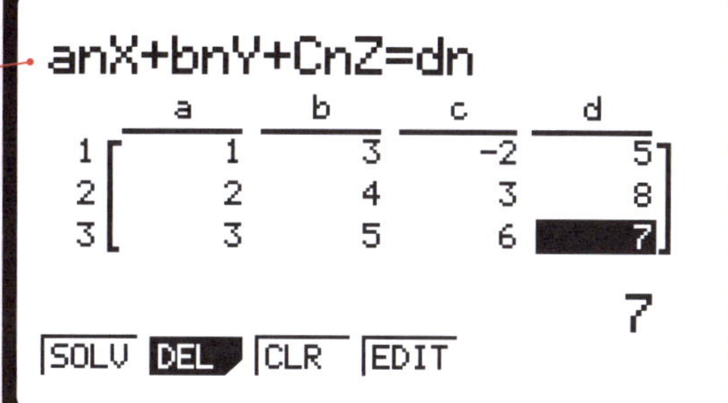

Example

$x + 3y - 2z = 5$ ①
$2x + 4y + 3z = 8$ ②
$3x + 5y + 6z = 7$ ③

→ **F1: Solve**

So the point of intersection of the lines is (-15, 8, 2).

Try this one:

$x = -5 - 2y - 3z$ ①
$3x + y - 3z = 4$ ②
$-3x + 4y + 7z = -7$ ③

This is a little trickier as you will need to rearrange before entering it into your calculator.

You should get a solution (-1, 1, -2).

Use a graphics calculator to solve these.

1

$$x + y - 1z = -1$$
$$2x - 2y + 3z = 8$$
$$2x - y + 2z = 9$$

2

$$2x + 3y - z = 24$$
$$x + y + 4z = 13$$
$$3x - 2y + z = -7$$

3

$$2x + 2y + 2z = 2$$
$$y + 3z = 5 - x$$
$$x + 4y + z = 10$$

4

$$3x + 2y + z = 20$$
$$4x - 10z = -10$$
$$-x - 2y + 2z = -1$$

5

$$x + 3y - z = 1$$
$$-2x - 6y + z = -3$$
$$3x + 5y = 4 + 2z$$

6

$$2x + 3y - 2z = 8$$
$$x - 4z = 1$$
$$-y - 6z + 2x = 4$$

7

$$x - 3y + 3z = -4$$
$$2x + 3y - z = 15$$
$$4x - 3y - z = 19$$

8

$$x = 2y + 1$$
$$z = -3x$$
$$x + 3y - z = 4$$

9

$$2x + 6y + 2z = -2$$
$$4x - 2y - 2z = 14$$
$$3x + 2z = 2y$$

10

$$x + 2y - 3z = 15$$
$$2x = 6 + 2z$$
$$x + z = 3$$

11

$$x + 2y - z = 23$$
$$3x + 8y - z = 179$$
$$-5x + 2y + 3z = 69$$

12

$$-20x - 10y - 30z = 30$$
$$100x - 150y + 50z = -650$$
$$120x - 180z = -660$$

 ISBN: 9780170389426

Equations without a unique solution

- So far we have dealt with **2 x 2** equations, which are **lines** meeting at a **point**, and **3 x 3** equations, which are **planes** meeting at a **point**.
- Meeting at a point with a unique solution is not the only possibility.
- At times you will find that you get **'Ma ERROR'** on your calculator. This will be for one of two reasons: the equations are either **inconsistent** or **dependent**.

Inconsistent equations: These have **no solutions**. The lines or planes **never meet**.

Dependent equations: These have **infinite solutions**. The lines or planes are **the same** as each other, but the equations may have been written in different ways.

2D equations where there is not a unique solution

1 Inconsistent equations

These have **no solutions**. The lines **never meet** at a point, so they are **parallel**.

Example: Solve the equations

$2x + 4y = 0$ ①

$x = 4 - 2y$ ②

Substitute ② into ①:

$$2x + 4y = 0$$
$$2(4 - 2y) + 4y = 0$$
$$8 - 4y + 4y = 0$$
$$8 = 0$$

This is a false statement!

If you plot these two equations you will notice that they are **parallel** lines. Therefore there is **no point of intersection**. This is why you don't get an answer.

Rearrangement of the equations into the form $y = mx + c$ makes this clear:

① $2x + 4y = 0$: $4y = -2x$ $\Rightarrow$ $1y = -\frac{1}{2}x$

② $x = 4 - 2y$: $2y + x = 4$ $\Rightarrow$ $1y = -\frac{1}{2}x + 2$

Remember that **m** represents the **gradient**. Both are $-\frac{1}{2}$.

Notice that if the equations have the same structure, the coefficients of x and y are in proportion but the constants are not:

① $2x + 4y = 0$: $2x + 4y = 0$

② $x = 4 - 2y$: $1x + 2y = 4$

$$\frac{2}{1} = \frac{4}{2} = 2 \neq \frac{0}{4}$$

Inconsistent equations have no solutions.
If you try to solve them, you will get a false statement.

ISBN: 9780170389426

2 Dependent equations

These have **infinite solutions**. The lines are **the same** as each other, but the equations have been written in different ways.

Example: Solve the equations

$$4x - 2y = 2 \quad ①$$
$$y = 2x - 1 \quad ②$$

Substitute ② into ①:

$$4x - 2y = 2$$
$$4x - 2(2x - 1) = 2$$
$$4x - 4x + 2 = 2$$
$$0x + 2 = 2$$
$$\mathbf{2 = 2}$$

This is a true statement but not very useful.

If you plot these two equations you will notice that they are the **same line** (equivalent). Therefore there is **no single point of intersection**. This is why you don't get an answer.

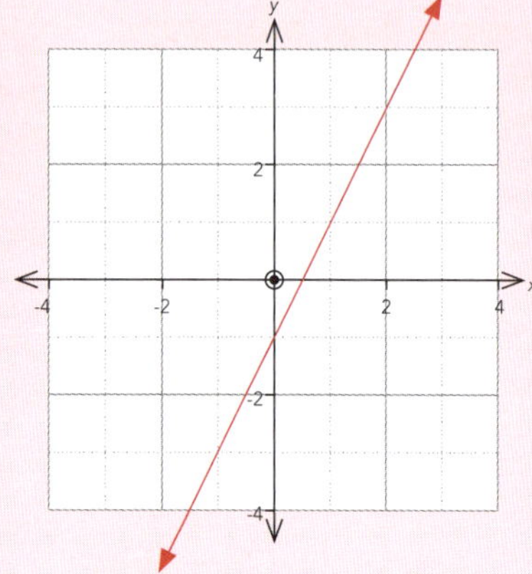

Rearrangement of the equations into the form $y = mx + c$ makes this clear:

② $y = 2x - 1$

① $4x - 2y = 2 \Rightarrow -2y = -4x + 2$

$\Rightarrow y = 2x - 1$

The same.

Notice that if the equations have the same structure, the coefficients of *x* and *y* and the constant are in proportion:

② $y = 2x - 1$: $\quad 1y = 2x - 1$

① $4x - 2y = 2$: $\quad 2y = 4x - 2$

$$\frac{1}{2} = \frac{2}{4} = \frac{-1}{-2}$$

Dependent equations have infinite solutions.
If you try to solve them, you will get a true but meaningless statement.

Example: Identify whether the equations in this set have a real solution, are inconsistent or are dependent: $x = 9 - 2y$ and $3x + 6y = 14$.

$$x = 9 - 2y \quad ①$$
$$3x + 6y = 14 \quad ②$$

Substitute for *x* in ②:

$$3(9 - 2y) + 6y = 14$$
$$27 - 6y + 6y = 14$$
$$\mathbf{27 = 14}$$

Untrue statement ⇒ equations are inconsistent and have no solutions.

 ISBN: 9780170389426

Identify whether these sets of equations have a real solution, are inconsistent or are dependent.

1

$x = 6 - y$
$2x + 2y = 12$

2

$3x + 3y = 15$
$6x = 18 - 4y$

3

$3x - y = 3$
$4x - 2y = 2$

4

$x + 5y = 10$
$2x + 10y = 30$

5

$8x + 2y = 18$
$y = 9 - 4x$

6

$2x - 3y = -2$
$-6x + 9y = 6$

7

$x - 3y - 12 = 0$
$y = \frac{1}{4}x - 7$

8

$y = -2x + 3$
$4x + 2y = 8$

ISBN: 9780170389426

9

$$2y + 6x = 16$$
$$y = -3x + 12$$

10

$$-2x + 2y = 4$$
$$x + y = 6$$

11

$$160x + 120y = 6$$
$$30y = -40x - 50$$

12

$$y = 3x + 2$$
$$6x - 2y + 4 = 0$$

13

$$3x + 3y = 15$$
$$2y = -2x + 6$$

14

$$y - 5x = 8$$
$$2x - 7y - 10 = 0$$

15 Find the value of k required to make this pair of equations inconsistent:

$$y = 2x - 7$$
$$kx - 2y = -2$$

16 Find the value of k required to make this pair of equations inconsistent:

$$4y = x + 20$$
$$3x = ky - 1.5$$

ISBN: 9780170389426

3D equations where there is not a unique solution

1 Inconsistent equations

These have **no solutions**. The planes **never meet** at a point.

Example: Solve the equations

$2x + 3y + 4z = 11$ ①
$x + 2y + z = 7$ ②
$x + y + 3z = 6$ ③

Eliminate x from ① and ②:

①: $2x + 3y + 4z = 11$
② x -2: $-2x - 4y - 2z = -14$ ④
① + ④: $-y + 2z = -3$ ⑤

Eliminate x from ② and ③:

②: $x + 2y + z = 7$
③: $x + y + 3z = 6$
② – ③: $y - 2z = 1$ ⑥

Solve ⑤ and ⑥:

$-y + 2z = -3$ ⑤
$y - 2z = 1$ ⑥
⑤ + ⑥: $\mathbf{0 = -2}$

This is a false statement!

A **false** statement shows:

- the three planes **never meet** at a point
- the equations are **inconsistent**, so there are **no solutions**.

There are three possible ways for this to occur:

1 All three planes are parallel to each other.

2 Two planes are parallel to each other, and the third is not.

3 None of the planes are parallel, but the lines where each pair of planes intersect are parallel to each other.

You need to be able to identify each of these situations, given a set of equations.

ISBN: 9780170389426

1 All three planes are parallel to each other.

Example:

$$x + y - z = 10 \quad ①$$
$$2x + 2y - 2z = 3 \quad ②$$
$$3x + 3y - 3z = 8 \quad ③$$

Eliminate x from ① and ②:

① x -2: $-2x - 2y + 2z = -20$ ④

②: $2x + 2y - 2z = 3$

④ + ②: **$0 = -17$**

False statement ⇒ ① and ② are inconsistent, and ∴ parallel.

Eliminate x from ① and ③:

① x -3: $-3x - 3y + 3z = -30$ ⑤

③: $3x + 3y - 3z = 8$

⑤ + ③: **$0 = -22$**

False statements.

False statement ⇒ ① and ③ are inconsistent, and ∴ parallel.

Because ① and ② are parallel, and ① and ③ are parallel:

- all **three** lines must be **parallel** to each other
- the system of equations is **inconsistent** so there are **no solutions**.

Notice that: 1 The **coefficients of x and y** are **in proportion**, but **the constants are not**:

$1x + 1y - 1z = 10$ ①

$2x + 2y - 2z = 3$ ②

$3x + 3y - 3z = 8$ ③

① and ②: $\frac{1}{2} = \frac{1}{2} = \frac{-1}{-2} \neq \frac{10}{3}$

② and ③: $\frac{2}{3} = \frac{2}{3} = \frac{-2}{-3} \neq \frac{3}{8}$

2 The equations are **multiples** of each other, **apart** from the **constants**:

$x + y - z = 10$

$2x + 2y - 2z = 3$ — **② = ① x 2**

$3x + 3y - 3z = 8$ — **③ = ① x 3**

Identification of this situation:

Coefficients of all three equations in proportion, but the constants are not in the same proportions as the coefficients.

 ISBN: 9780170389426

2 Two planes are parallel to each other, and the third is not.

Example:

$$x - 2y + 3z = 12 \quad ①$$
$$3x - y + 2z = 9 \quad ②$$
$$-2x + 4y - 6z = -15 \quad ③$$

Eliminate x from ① and ②:

① x -3: $-3x + 6y - 9z = -36$ ④

②: $3x - y + 2z = 9$

④ + ②: $5y - 7z = -27$ ⑤

Because $5y - 7z = -25$ represents a line, ① and ② are not parallel.

Eliminate x from ① and ③:

① x 2: $2x - 4y + 6z = 24$ ⑤

③: $-2x + 4y - 6z = -15$

⑤ + ③: **$0 = 9$** — False statement.

False statement ⇒ ① and ③ are inconsistent, and ∴ parallel.

Because ① and ③ are parallel, and ① and ② are not parallel:

- **two** lines must be **parallel** to each other, and the third is not.
- the system of equations is **inconsistent** — there are **no solutions**.

Notice that: 1 Equations ① and ③ have the same structure and the **coefficients of x and y** are **in proportion,** but **the constants are not** ⇒ lines are **parallel**:

$$1x - 2y + 3z = 12 \quad ①$$
$$-2x + 4y - 6z = -15 \quad ③$$

$$\frac{1}{-2} = \frac{-2}{4} = \frac{3}{-6} \neq \frac{12}{-15}$$

Equations ① and ② have **coefficients of x and y** that are **not in proportion** ⇒ lines **not parallel:**

$$1x - 2y + 3z = 12 \quad ①$$
$$3x - 1y + 2z = 9 \quad ②$$

$$\frac{1}{3} \neq \frac{-2}{-1} \neq \frac{3}{2}$$

2 Two equations are **multiples** of each other, **apart** from the **constants**:

$$x - 2y + 3z = 12 \quad ①$$
$$3x - y + 2z = 9 \quad ②$$
$$-2x + 4y - 6z = -15 \quad ③$$

② — **Not a multiple of ① or ③.**

③ — **③ = ① x -2**

Identification of this situation:

Coefficients of only two equations in proportion, but the constants are not in the same proportions as the coefficients.

3 No planes are parallel, but they intersect along lines that are parallel to each other.

Example:

$3x - y - 5z = 19$ ①
$2x + 3y - 7z = 6$ ②
$x - 4y + 2z = -7$ ③

Eliminate y from ① and ②:

① x 3: $9x - 3y - 15z = 57$ ④
②: $2x + 3y - 7z = 6$
④ + ②: $11x - 22z = 63$ ⑤

Eliminate y from ① and ③:

① x -4: $-12x + 4y + 20z = -76$ ⑥
③: $x - 4y + 2z = -7$
⑥ + ③: $-11x + 22z = -83$ ⑦

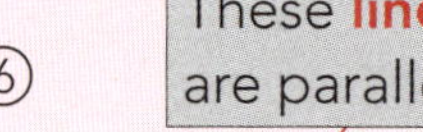

Solve ⑤ and ⑦:

⑤: $11x - 22z = 63$
⑦ $-11x + 22z = -83$
⑤ + ⑦: $0 = -20$

This is a false statement!

Because equations ⑤ and ⑦ are for parallel but distinct lines:

- the planes are **not parallel**
- the system of equations is **inconsistent** — there are **no solutions**
- the planes meet in **lines** that are **parallel with each other**.

Notice that:

$3x - 1y - 5z = 19$ ①
$2x + 3y - 7z = 6$ ②
$1x - 4y + 2z = -7$ ③

② + ③ parallel with ①.

Identification of this situation:

A linear combination of any two equations produces a plane parallel with the third plane.

ISBN: 9780170389426

Show why these equations are inconsistent, determine the relationships between the equations, and show what the planes would look like.

1

$$x + 3y + 6z = 10$$
$$2x - 4y - 5z = 16$$
$$3x - y + z = 5$$

What would these planes look like?

2

$$x + 2y + 3z = 4$$
$$x + 5y + 7z = 8$$
$$2x + 4y + 6z = 9$$

What would these planes look like?

3

$$4x + 2y + 6z = 3$$
$$2x + y + 3z = 16$$
$$6x + 3y + 9z = 9$$

What would these planes look like?

4

$$3x + 2y + 4z = 1$$
$$x - y - 2z = 3$$
$$2x + 3y + 6z = 8$$

What would these planes look like?

ISBN: 9780170389426

2 Dependent equations

- These have infinite **solutions**.
- If you try to solve them, you will get a **true** but **meaningless** statement.

There are three possible ways for this to occur:

1 All three planes are the same.

Example: Solve the equations

$$x - y + z = 3 \quad ①$$
$$2x - 2y + 2z = 6 \quad ②$$
$$3x - 3y + 3z = 9 \quad ③$$

Eliminate x from ① and ②:

① x 2: $2x - 2y + 2z = 6$ ④
②: $2x - 2y + 2z = 6$
④ – ②: $0 = 0$

This true statement means that planes ① and ② are the same.

Eliminate x from ① and ③:

① x 3: $3x - 3y + 3z = 9$ ⑤
③: $3x - 3y + 3z = 9$
⑤ – ③: $0 = 0$

This true statement means that planes ① and ③ are the same.

Because ① and ② are the same plane, and ① and ③ are the same plane:

- all **three** planes are the **same**
- the system of equations is **dependent** so there are **infinite solutions**.

Notice that: 1 The coefficients of x, y and z **and the constants** are **in proportion**:

$$1x - 1y + 1z = 3 \quad ①$$
$$2x - 2y + 2z = 6 \quad ②$$
$$3x - 3y + 3z = 9 \quad ③$$

① and ②: $\frac{1}{2} = \frac{-1}{-2} = \frac{1}{2} = \frac{3}{6}$

② and ③: $\frac{2}{3} = \frac{-2}{-3} = \frac{2}{3} = \frac{6}{9}$

2 The equations are **multiples** of each other:

$$x - y + z = 3$$
$$2x - 2y + 2z = 6$$
$$3x - 3y + 3z = 9$$

② = ① x 2

③ = ① x 3

Identification of this situation:

Equations are multiples of each other.

 ISBN: 9780170389426

2 Two planes are the same and the third is not parallel to either.

Example: Solve the equations

$$3x - y + 2z = 9 \quad ①$$
$$6x - 2y + 4z = 18 \quad ②$$
$$x + 3y - 5z = 0 \quad ③$$

Eliminate x from ① and ②:

① x -2: $-6x + 2y - 4z = -18$ ④

②: $6x - 2y + 4z = 18$

④ + ②: $0 = 0$

This true statement means that planes ① and ② are the same: ② = 2 x ①.

Equations ① and ③ have **coefficients of x, y and z** that are **not in proportion** ⇒ these planes **not parallel**.

$3x - y + 2z = 9$ ①

$x + 3y - 5z = 0$ ③

$$\frac{3}{1} \neq \frac{-1}{3} \neq \frac{2}{-5}$$

This confirms that planes ① and ② are the same planes, but plane ③ is not parallel to ① and ②.

Because ① and ② are the same plane, and ③ is not parallel to ① and ②:

- plane ③ must intersect planes ① and ② in a line
- the system of equations is **dependent** so there are **infinite solutions** along the line.

Notice that:

1 The coefficients of x, y and z **and the constants** are **in proportion** for ① and ②, but not ③ :

$3x - 1y + 2z = 9$ ①

$6x - 2y + 4z = 18$ ②

$1x + 3y - 5z = 0$ ③

$$\frac{3}{6} = \frac{-1}{-2} = \frac{2}{4} = \frac{9}{18}$$

$$\frac{6}{1} \neq \frac{-2}{3} \neq \frac{4}{-5} \neq \frac{18}{0}$$

2 Only **two** equations are **multiples** of each other:

$3x - y + 2z = 9$ ①

$6x - 2y + 4z = 18$ ② — **② = ① x 2**

$x + 3y - 5z = 0$ ③ — **Not a multiple of ① or ②.**

Identification of this situation:

Two equations are multiples of each other, the third is not.

ISBN: 9780170389426

3 The three planes meet along a line — like the leaves of a book meeting along the spine.

Example: Solve the equations:

$$4x - y + z = 5 \quad ①$$
$$2x - 2y + 3z = 7 \quad ②$$
$$6x - 3y + 4z = 12 \quad ③$$

Eliminate y from ① and ②:

① x -2: $-8x + 2y - 2z = -10$ ④

②: $2x - 2y + 3z = 7$

④ + ②: $-6x + z = -3$ ⑤

Eliminate y from ① and ③:

① x -3: $-12x + 3y - 3z = -15$ ⑥

③: $6x - 3y + 4z = 12$

⑥ + ③: $-6x + z = -3$ ⑦

Solve ⑤ and ⑦:

⑤ x -1: $6x - z = 3$ ⑧

⑦ $-6x + z = -3$

⑧ + ⑦: $0 = 0$ — True statement.

Because a true statement results from solving ⑦ and ⑧, the **two lines** are the **same**, so:

- all **three** planes must meet along the **same line**
- the system of equations is **dependent** so there are **infinite solutions**.

Notice that: 1 The coefficients of x, y and z are **not in proportion**:

$$4x - 1y + 1z = 5 \quad ①$$
$$2x - 2y + 3z = 7 \quad ②$$
$$6x - 3y + 4z = 12 \quad ③$$

① and ②: $\frac{4}{2} \neq \frac{-1}{-2} \neq \frac{1}{3}$

② and ③: $\frac{2}{6} \neq \frac{-2}{-3} \neq \frac{3}{4}$

2 Two equations can be added to form the third:

$$4x - y + z = 5 \quad ①$$
$$2x - 2y + 3z = 7 \quad ②$$
$$6x - 3y + 4z = 12 \quad ③$$

③ = ① + ②

This is a **linear combination**. These may involve multiplication by constants.

Identification of this situation:

A linear combination of any two equations produces the third equation, but the coefficients are not in proportion.

 ISBN: 9780170389426

In trickier examples, linear combinations may involve multiplication by constants.

Example: Solve the equations

$3x - 2y + z = 5$ ①
$1x + 4y - 2z = 8$ ②
$11x + 2y - z = 31$ ③

In this case, simply adding or subtracting the equations does not work.

Method 1: Solving by hand

Suppose: a x ① + b x ② = ③

Then:
1. Consider the x terms: $3a + 1b = 11$ ④
2. Consider the y terms: $-2a + 4b = 2$ ⑤

Solve these simultaneously:

2 x ④: $6a + 2b = 22$
3 x ⑤: $-6a + 12b = 6$ +
$14b = 28$
$b = 2$
$\therefore a = 3$

So: **3** x ① + **2** x ② = ③

Method 2: Solving on your calculator

$3x - 2y + z = 5$ ①
$1x + 4y - 2z = 8$ ②
$11x + 2y - z = 31$ ③

Enter these coefficients into your graphics calculator, but going **across**:

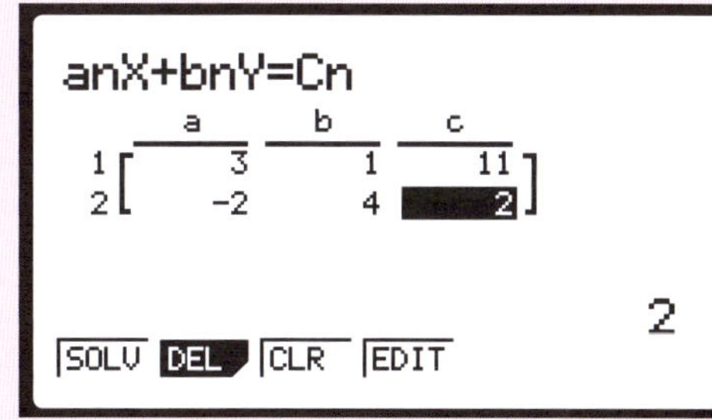

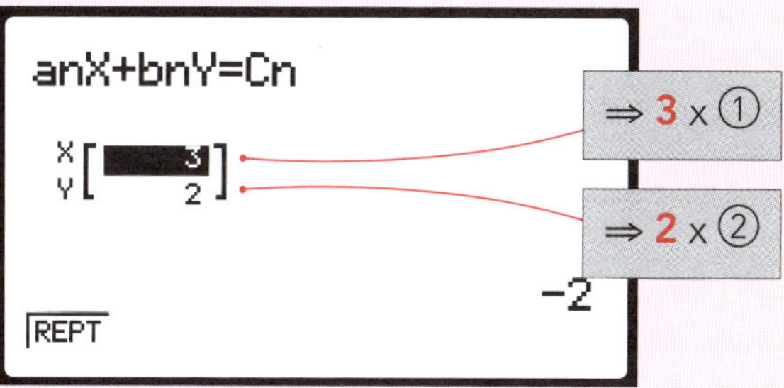

So: **3** x ① + **2** x ② = ③

Whichever method you use, check your answer.

Check:

3 x ①: $9x - 6y + 3z = 15$
2 x ②: $2x + 8y - 4z = 16$ +
$11x + 2y - z = 31$ = ③ ✓

ISBN: 9780170389426

Show why these equations are dependent, determine the relationships between the equations, and show what the planes would look like.

1

$$x + 2y + 3z = 4$$
$$x + y - 2z = 3$$
$$2x + 4y + 6z = 8$$

What would these equations look like?

2

$$7x - 5y + 3z = 4$$
$$8x - 2y + z = 13$$
$$x + 3y - 2z = 9$$

What would these equations look like?

3

$$x - 3y + z = 4$$
$$-x + 4y - 3z = 5$$
$$x - 2y - z = 13$$

What would these equations look like?

4

$$x - 2y + z = 1$$
$$2x - 5y + 3z = 4$$
$$2x - 3y + z = 0$$

What would these equations look like?

ISBN: 9780170389426

5

$$x - 2y - z = 3$$
$$2x - 4y - 2z = 6$$
$$4x - 8y - 4z = 12$$

What would these equations look like?

6

$$x + 3y - 2z = 4$$
$$-2x - y + z = -2$$
$$-x + 7y - 4z = 8$$

What would these equations look like?

7

$$x - 3y + 3z = 2$$
$$-2x + 2y - 4z = 1$$
$$x + 9y - 3z = -13$$

What would these equations look like?

8

$$2x + y - 2z = 5$$
$$3x + 2y + 5z = 5$$
$$4x + 2y - 4z = 10$$

What would these equations look like?

ISBN: 9780170389426

Finding multiple solutions for dependent equations

Always keep in mind the context: in most situations, answers must be integral (whole numbers), and cannot be negative.

Example 1: Find multiple solutions to the equations

$$3x - y + 2z = 9 \quad ①$$
$$6x - 2y + 4z = 18 \quad ②$$
$$x + 3y - 5z = 0 \quad ③$$

From page 33: Planes ① and ② are the same (② = 2 x ①), and ③ is not parallel.
∴ Solutions lie along the line where ① (or ②) and ③ meet.

Solve ① and ③:

$$3x - y + 2z = 9 \quad ①$$
$$x + 3y - 5z = 0 \quad ③$$

You must use the line that is different from the other two (③).

① x 3: $9x - 3y + 6z = 27$
③: $x + 3y - 5z = 0$

$$10x + z = 27$$

$$\therefore \mathbf{z = 27 - 10x}$$

Possible values for x and z:

1 $x = 0, z = 27$: Substitute into ③: $0 + 3y - 5(27) = 0$
$$y = \frac{5(27) - 0}{3} = 45$$
∴ Solution is (0, 45, 27).

2 $x = 1, z = 17$: Substitute into ③: $1 + 3y - 5(17) = 0$
$$y = \frac{5(17) - 1}{3} = 28$$
∴ Solution is (1, 28, 17).

3 $x = 2, z = 7$: Substitute into ③: $2 + 3y - 5(7) = 0$
$$y = \frac{5(7) - 2}{3} = 11$$
∴ Solution is (2, 11, 7).

Notice:
In this situation there are no other integral (whole number) positive solutions because if $x = 3$, $z = -3$, and if x is 3 or more, the values of z will all be negative. Negative values are very unlikely in practical situations.

ISBN: 9780170389426

Cut out each of the cards, mix the pieces, and assemble seven complete sets, based on the diagrams.

Picture			
Identification	Can be solved by hand or on your calculator. A	Three equations: Coefficients in proportion. Constants not in proportion. S	Two equations: Coefficients in proportion. Constants not in proportion. Third equation — not in proportion. F
Result when you try to solve the system	Get one solution V	False statement e.g. $5 = 0$ K	False statement e.g. $5 = 0$ K
Example	$x + 2y - 4z = 1$ $3x + y - 4z = 7$ $2x + 6y - 8z = 10$ E	$x + 3y - 4z = 6$ $3x + 9y - 12z = 12$ $2x + 6y - 8z = 10$ U	$x + 3y - 4z = 2$ $3x + y - z = 5$ $2x + 6y - 8z = 10$ T
Solutions (if any)	(5, 4, 3) N	None X	None X

ISBN: 9780170389426

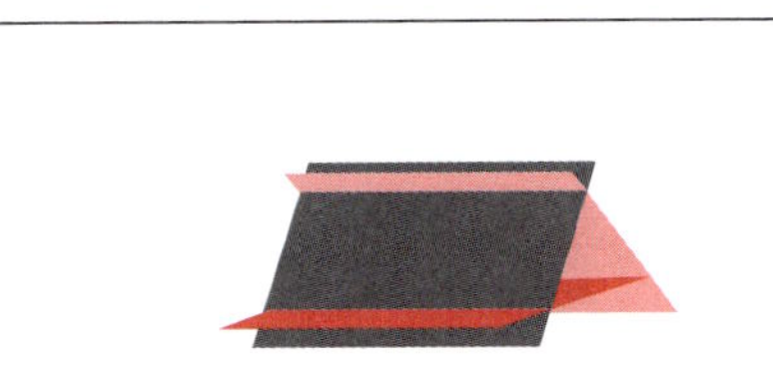			
Linear combinations of each other. e.g. ③ = ① + ② But not for constants M	Coefficients and constants in proportion. W	Two equations: Coefficients and constants in proportion. Third equation — not in proportion. C	Linear combinations of each other. e.g. ③ = ① + ② P
False statement e.g. $5 = 0$ K	True statement e.g. $2 = 2$ B	True statement e.g. $2 = 2$ B	True statement e.g. $2 = 2$ B
$x + 4y - 3z = 5$ $x + 2y - 5z = 4$ $2x + 6y - 8z = 10$ G	$x + 3y - 4z = 5$ $3x + 9y - 12z = 15$ $2x + 6y - 8z = 10$ R	$x + 3y - 4z = 5$ $x + 2y - 3z = 3$ $2x + 6y - 8z = 10$ D	$x + 4y - 3z = 6$ $x + 2y - 5z = 4$ $2x + 6y - 8z = 10$ H
None X	Infinite e.g. (0, 3, 1), (0, 7, 4), (1, 4, 2), (1, 8, 5), (2, 5, 3), (3, 6, 4), …. L	Infinite e.g. (0, 3, 1), (1, 4, 2), (2, 5, 3), (3, 6, 4), …. Z	Infinite, but the only non-negative, integral answers are (2, 1, 0), (9, 0, 1). J

ISBN: 9780170389426

Always keep in mind the context: in most situations, answers must be integral (whole numbers), and cannot be negative.

Example 2: Find multiple solutions to the equations

$$4x - y + z = 5 \quad ①$$
$$2x - 2y + 3z = 7 \quad ②$$
$$6x - 3y + 4z = 12 \quad ③$$

From page 34: Planes ①, ② and ③ meet along a line because ① + ② = ③.
∴ Solutions lie along the line where ①, ② and ③ meet.

Solve ① and ②:

$$4x - y + z = 5 \quad ①$$
$$2x - 2y + 3z = 7 \quad ②$$

You could use any two of the three equations, but select the easiest.

① x -2: $-8x + 2y - 2z = -10$
②: $2x - 2y + 3z = 7$

$$-6x + z = -3$$

$$\therefore \mathbf{z = 6x - 3}$$

Possible values for *x* and *z*:

1 $x = 1, z = 3$: Substitute into ①: $4(1) - y + 3 = 5$
$y = 2$
∴ Solution is (1, 2, 3).

2 $x = 2, z = 9$: Substitute into ①: $4(2) - y + 9 = 5$
$y = 12$
∴ Solution is (2, 12, 9).

3 $x = 3, z = 15$: Substitute into ①: $4(3) - y + 15 = 5$
$y = 22$
∴ Solution is (3, 22, 15).

Notice:
- There will be many more solutions in this case because as *x* increases, so do *y* and *z*.
- There is also a pattern of increase because all the points lie along a line.

Find multiple solutions for the following sets of equations.

1

$$x + 2y + 3z = 40 \quad (1)$$
$$2x + 4y + 6z = 80 \quad (2)$$
$$x + y + 2z = 25 \quad (3)$$

2

$$x + 2y + z = 23 \quad (1)$$
$$2x + 5y + 3z = 59 \quad (2)$$
$$2x + 3y + z = 33 \quad (3)$$

3

$$x + 3y + z = 31 \quad (1)$$
$$x + 4y + 2z = 46 \quad (2)$$
$$3x + 10y + 4z = 108 \quad (3)$$

4

$$x + y + 2z = 20 \quad (1)$$
$$2x + 4y + 5z = 50 \quad (2)$$
$$x + 3y + 3z = 30 \quad (3)$$

ISBN: 9780170389426

5

$$5x + 2y + z = 26 \quad ①$$
$$11x + 4y + z = 48 \quad ②$$
$$15x + 6y + 3z = 78 \quad ③$$

6

$$2x + y + 2z = 33 \quad ①$$
$$3x + 2y + 5z = 67 \quad ②$$
$$4x + 2y + 4z = 66 \quad ③$$

7

$$3x + y + 2z = 28 \quad ①$$
$$x + 2y + 2z = 19 \quad ②$$
$$4x + 3y + 4z = 47 \quad ③$$

8

$$7x - 5y + 3z = 59 \quad ①$$
$$8x - 2y + z = 75 \quad ②$$
$$x + 3y - 2z = 16 \quad ③$$

ISBN: 9780170389426

3D summary

Solutions	Result when solving	What the planes do	Recognition from equations
Unique solution	(x, y, z)	All three intersect	Can be solved by hand or on your calculator.
No solutions	False statement e.g. 5 = 0	All three parallel	Three equations: Coefficients in proportion. Constants not in proportion. e.g. $\frac{1}{2} = \frac{1}{2} = \frac{-1}{-2} \neq \frac{10}{3}$
		Two parallel, one not	Two equations: Coefficients in proportion. Constants not in proportion. Third equation — not in proportion.
		Lines of intersection parallel (tent)	Variables are linear combinations of each other. e.g. ① = ② + ③ But not the constants.
Infinite solutions	True statement e.g. 2 = 2	All three planes the same	Coefficients and constants in proportion.
		Two planes the same, third not parallel	Two equations: Coefficients and constants in proportion. Third equation — not in proportion.
		All three planes meet in a line (pages of a book)	Equations are linear combinations of each other. e.g. ① = ② + ③

ISBN: 9780170389426

Putting it together

Work out whether these sets of equations have a unique solution, are inconsistent or dependent. Determine the relationships between the equations and show what the planes would look like.

1

$$3x - 5y + 12z = -6$$
$$2x - 2y + 6z = 10$$
$$4x - 4y + 12z = 18$$

What would these equations look like?

2

$$2x + 2y + 2z = 2$$
$$x + y + 3z = 5$$
$$x + 4y + z = 10$$

What would these equations look like?

3

$$x - 3y - 4z = 3$$
$$3x + 4y - z = 13$$
$$2x - 19y - 19z = 2$$

What would these equations look like?

4

$$-2x - 4y - z = 23$$
$$-x + 3y - 2z = 14$$
$$3x + y + 3z = 19$$

What would these equations look like?

ISBN: 9780170389426

5

$$-3x + 6y - 9z = -1$$
$$3x - 4y + 6z = 2$$
$$4x - 8y + 12z = 3$$

What would these equations look like?

6

$$20x + 10y + 35z = 255$$
$$32x + 16y + 30z = 266$$
$$16x + 8y + 28z = 204$$

What would these equations look like?

7

$$5x = -2y - 3z - 5$$
$$3x + 2z = 9$$
$$-4x + 2y - 3z = 8$$

What would these equations look like?

8

$$3x - 3y - 6z = -3$$
$$2x - 2y - 4z = 10$$
$$-2x + 3y + z = 7$$

What would these equations look like?

ISBN: 9780170389426

9

$$2x + 3y + z = 5$$
$$3x + 7y - 36z = -25$$
$$x + 2y - 7z = -4$$

What would these equations look like?

10

$$x + 2y + z = 10$$
$$3y - 3x = 27 + 5z$$
$$-y = -5 - 2x - 3z$$

What would these equations look like?

11

$$x - 4y + 2z = -2$$
$$x + 2y - 2z = -3$$
$$x - y = 4$$

What would these equations look like?

12

$$2x + y - 2z = 5$$
$$3x + 2y + 5z = 5$$
$$4x + 2y - 4z = 10$$

What would these equations look like?

Forming and solving simultaneous equations

1 Forming and solving 2 x 2 simultaneous equations

Hints:

- Read what the question asks for — this will tell you what your variables will be.
- Define your variables using their first letters.

Examples:

1 Tom has twice as many marbles as Emily. If Emily had another 25 marbles, she would have had three times as many as Tom. **How many marbles does each person have?**

> Your variables will need to be the **numbers of marbles that each person had**.

Call the number of marbles Tom had **t**.
Call the number of marbles Emily had **e**.

> Always define your **variables**.

$t = 2e$ ①
$e + 25 = 3t$ ②

Substitute ① into ②: $e + 25 = 3(2e)$
$e + 25 = 6e$ $(-e)$
$25 = 5e$
$\mathbf{e = 5}$

Substitute for e in ①: $t = 2(5)$
$\mathbf{t = 10}$

Check: ① $10 = 2(5)$ ✓
② $5 + 25 = 3(10)$ ✓

∴ **Tom had 10 marbles and Emily had 5.**

> Always answer the question in a **sentence**.

2 Nick buys six peaches and five apples and they cost him \$5.70.
Four peaches and seven apples cost Katie \$4.90.
Calculate the **prices of peaches and apples.**
Call the price of peaches **p**.
Call the price of apples **a**.

> Your variables will need to be the **prices of peaches and apples**.

$6p + 5a = 5.70$ ①
$4p + 7a = 4.90$ ②

Multiply ① by -7: $-42p - 35a = -39.90$ ③
Multiply ② by 5: $20p + 35a = 24.50$ ④
Add ③ and ④: $-22p = -15.4$
$\mathbf{p = 0.70}$

Substitute for p in ①: $6(0.70) + 5a = 5.70$
$5a = 5.70 - 4.20$
$5a = 1.50$
$\mathbf{a = 0.30}$

Check: ① $6(0.70) + 5(0.30) = 5.70$ ✓
② $4(0.70) + 7(0.30) = 4.90$ ✓

∴ **Peaches cost \$0.70 and apples cost \$0.30.**

ISBN: 9780170389426

Form and solve these equations.

1 The difference of two numbers is two. Their sum is 14. Find the values of the two numbers.

2 In a furniture store on Saturday, they sold five chairs and seven tables and earned $1790. On Sunday, they sold one chair and one table and earned $290. How much did they sell each item for?

3 Circus tickets cost $4 for children and $7 for adults. They sell 517 tickets altogether and make $2617. How many of each ticket were sold?

4 Julia and Angus have $35 between them. Julia has $4 less than twice what Angus has. How much money does each have?

5 Three ice creams and three drinks cost $17.10. Your friend bought four ice creams and two drinks, which cost a total of $17.80. How much do ice creams cost?

6 Kora and Jo have a combined age of 48. Four years ago, Kora was three times the age Jo is now. How old are they now?

ISBN: 9780170389426

7 2 kg of apples and 6 kg of carrots cost \$19; 1 kg of apples and 5 kg of carrots cost \$15. How much do carrots and apples cost per kilogram?

8 Jeremy thinks of two numbers. When he doubles the first and adds the second, the result is 17. When the first is tripled and the second is subtracted from the answer, the result is 18. What are the two numbers Jeremy started with?

9 The Tall Ferns scored a total of 80 points in their game. They made a total of 34 baskets. How many two-point baskets and three-point baskets did they make?

10 Henry bought seven tops and five pairs of pants for a total of \$364. Fetu spent \$550 on 10 tops and eight pairs of pants. How much were the tops and pants?

11 Sarah and Charlie's dinner cost \$45 altogether. Sarah's meal cost \$6 more than Charlie's. How much should each pay?

12 A total of 23 beetles and spiders lived under a log. If there were 152 legs (and none had lost legs), how many spiders were there?

ISBN: 9780170389426

2 Forming and solving 3 x 3 simultaneous equations

Don't forget:

- Read what the question asks for — this will tell you what your variables will be.
- Define your variables using their first letters.

Example:

120 people went to the movies. Adults' tickets cost \$15, children's cost \$12 and seniors' tickets cost \$10 each. After the movie session, the cashier had a total of \$1515. She had sold twice as many children's tickets as adults' tickets. **How many of each type of ticket** did she sell?

Your variables will need to be the **numbers of each type** of ticket sold.

Call the number of adults' tickets **a**.

Call the number of children's tickets **c**.

Call the number of seniors' tickets **s**.

Always define your **variables**.

$$c = 2a \quad ①$$
$$a + c + s = 120 \quad ②$$
$$15a + 12c + 10s = 1515 \quad ③$$

Substitute ① into ②:

$$a + c + s = 120$$
$$a + (2a) + s = 120$$
$$\mathbf{3a + s = 120} \quad ④$$

Substitute ① into ③:

$$15a + 12c + 10s = 1515$$
$$15a + 12(2a) + 10s = 1515$$
$$\mathbf{39a + 10s = 1515} \quad ⑤$$

④ x 10: $30a + 10s = 1200$ ⑥

⑤ – ⑥: $9a = 315$

$$\mathbf{a = 35}$$

Substitute for **a** into ①: $\mathbf{c = 70}$

Substitute for **a** and **c** in ②: $35 + 70 + s = 120$

$$\mathbf{s = 15}$$

Check:

① : $c = 2(35) = 70$ ✓

② : $35 + 70 + 15 = 120$ ✓

③ : $15(35) + 12(70) + 10(15) = 1515$ ✓

It's a good idea to make sure all three equations work with all three values.

∴ **There were 35 adults' tickets sold, 70 children's tickets and 15 seniors' tickets.**

Always answer the question in a **sentence**.

ISBN: 9780170389426

Write three simultaneous equations and then solve to find the solution.

1 Three friends go to the market. Chris bought 2 kg apples, 1.5 kg broccoli and 3 kg carrots, which cost a total of $14.50. Pina bought 1 kg apples, 1 kg broccoli and 4 kg carrots, which cost a total of $12. Gloria bought 2.5 kg apples, 3 kg broccoli and 1 kg carrots, which cost a total of $15.
Calculate the cost per kilogram of apples, broccoli and carrots.

2 Lelei sells bags of fudge. She sells a small bag for $3, a medium bag for $5 and a large bag for $7. In total she sells 15 bags and makes $83. She sells two more medium bags than small bags. How many of each size did she sell?

3 The PE department has ordered some new basketball equipment. They need basketballs, cones and nets for the hoops. Eight basketballs, 12 cones and four nets cost them $415.76.
Ten cones cost $13.09 less than a basketball. One cone cost $6 less than a net. Calculate the cost of each.

4 Anna, Betty and Carlos altogether have $570. Anna has $60 more than Betty. Carlos has $30 less than Anna. How much does each have?

  ISBN: 9780170389426

5 Sefa received 109 text messages in the last three days. On Monday he received six fewer messages than on Wednesday. On Tuesday he received three times as many as on Wednesday. How many text messages did he get on each day?

6 The new strip for the school basketball team costs $112, and consists of a pair of shorts, a singlet and a tracksuit. The singlets cost $4 more than a pair of shorts. For the price of a tracksuit you could buy five pairs of shorts and have $3 left over. Calculate the prices of shorts, singlets and tracksuits.

7 Kristen, Erin and Hugh have played 54 games of Labyrinth. If Erin had won 12 more games, she would have won twice as many as Hugh. Kristen has won six more games than Hugh. How many games did each of them win?

8 Mason, Xavier and Archer have a combined birthday party. On the cake there are 88 candles to represent their combined ages. Xavier has 10 times as many candles as Archer. Archer has 16 fewer candles than Mason. How many candles does each of them have?

ISBN: 9780170389426

9 The largest angle of a triangle is 12° less than the sum of the other two angles, and it is 12° more than four times the smallest angle. Calculate the sizes of the angles.

10 The sum of the digits of a three-digit number is 15. The 100s digit is three times the 10s digit. If the number is reversed, the new number is 99 more than the original number. Use simultaneous equations to find the original three-digit number.

11 The general equation for a parabola can be written as $y = ax^2 + bx + c$. A parabola passes through three points: (2, 9), (-3, 34) and (5, 42). Use simultaneous equations to calculate the values of a, b and c. Write the equation of the parabola.

12 Find the equation of a cubic that passes through the following points: (0, 5), (-2, 17), (1, 8) and (-4, 13).

ISBN: 9780170389426

Applications: what do I do?

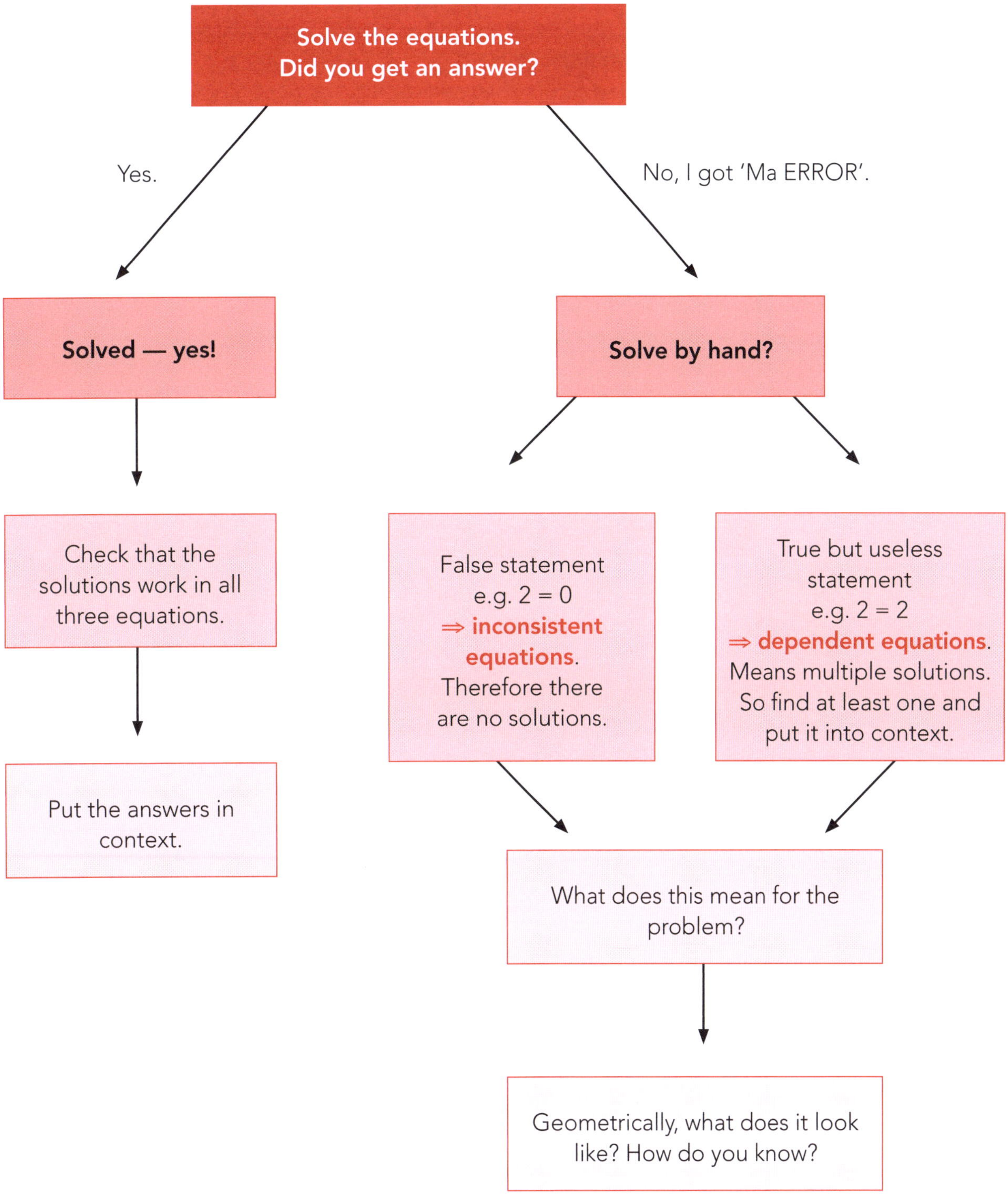

Solve the equations. Did you get an answer?

Yes.

Solved — yes!

Check that the solutions work in all three equations.

Put the answers in context.

No, I got 'Ma ERROR'.

Solve by hand?

False statement e.g. $2 = 0$ ⇒ **inconsistent equations**. Therefore there are no solutions.

True but useless statement e.g. $2 = 2$ ⇒ **dependent equations**. Means multiple solutions. So find at least one and put it into context.

What does this mean for the problem?

Geometrically, what does it look like? How do you know?

Applications involving inconsistent and dependent equations

Don't forget:
- Read what the question asks for — this will tell you what your variables will be.
- Define your variables using their first letters.

If the equations are inconsistent:
- Describe or draw what they look like.
- Explain what this means for the problem.

If the equations are dependent:
- Describe or draw what they look like.
- Explain what this means for the problem.
- Find at least one example of a solution, and put it into the context.

Example: The school council is making muffins to sell on Gala Day. George, Maggie and Eru have different recipes. They have donations of 95 eggs, 9.8 kg of cheese and 13.9 kg of flour, and they want to make as many complete batches as possible while using up all these ingredients.

	Eggs	Cheese (g)	Flour (g)
George	4	500	550
Maggie	2	300	300
Eru	3	200	450

a Write equations for these relationships, and solve them to find the number of batches that should be made for each recipe.

$$4g + 2m + 3e = 95 \quad ①$$
$$500g + 300m + 200e = 9800 \quad ②$$
$$550g + 300m + 450e = 13\,900 \quad ③$$

$$\therefore g = 7, m = 11 \text{ and } e = 15$$

Answer in a sentence.

So they should make 7 batches of George's recipe, 11 batches of Maggie's and 15 batches of Eru's recipe.

b George remembered that last time he made this recipe, it was better with 600 g of flour, rather than 550. Investigate what difference this makes to their solutions to the equations, and describe how the three planes made by the equations relate to each other.

New equations:

$$4g + 2m + 3e = 95 \quad ①$$
$$500g + 300m + 200e = 9800 \quad ②$$
$$600g + 300m + 450e = 13\,900 \quad ③$$

 ISBN: 9780170389426

For equations ① and ③, the coefficients are proportional but the constants are not

$\left(\frac{4}{600} = \frac{2}{300} = \frac{3}{450} \neq \frac{95}{13\,900}\right)$. *Apart from the constants, equation ③ = ① x 150. This means that planes ① and ③ are parallel.*

Plane ② is not parallel to either of the other planes because its coefficients are not in proportion with those of either ① or ③.

Geometrically:

∴ There are no solutions to this system of equations because the three planes never meet. This means that they will not be able to make complete batches of muffins that use up all the ingredients.

c The foods teacher said she would donate another 350 g of flour. Assume that they used George's better recipe with 600 g of flour and they want to use up all the donated ingredients. Investigate what difference this makes to their solutions to the equations, and describe how the three planes made by these equations relate to each other.

New equations:

$$4g + 2m + 3e = 95 \quad ①$$
$$500g + 300m + 200e = 9800 \quad ②$$
$$\mathbf{600}g + 300m + 450e = \mathbf{14\,250} \quad ③$$

Once again, for equations ① and ③, the coefficients are proportional but this time the constants are also in proportion $\left(\frac{4}{600} = \frac{2}{300} = \frac{3}{450} = \frac{95}{14\,250} = \frac{1}{150}\right)$.

Equation ③ = ① x 150.

This means that planes ① and ③ are the same, but plane ② is not, so the three planes meet along a line, and there are many solutions to this system of equations along the line.

Geometrically:

To find some solutions:

③ – ②: $100g + 250e = 4450$ ④

④ ÷ 50: $2g + 5e = 89$

$$\therefore g = \frac{89 - 5e}{2}$$

Note that e must be odd for integral answers.

From ①: $m = \frac{95 - 4g - 3e}{2}$

Let e = 1: $g = \frac{89 - 5(1)}{2} = 42 \Rightarrow m = \frac{95 - 4(42) - 3(1)}{2} = -38$

Let e = 11: $g = \frac{89 - 5(11)}{2} = 17 \Rightarrow m = \frac{95 - 4(17) - 3(11)}{2} = -3$

Let e = 13: $g = \frac{89 - 5(13)}{2} = 12 \Rightarrow m = \frac{95 - 4(12) - 3(13)}{2} = 4$

⇒ Solution is (12, 4, 13).

Let e = 15: $g = \frac{89 - 5(15)}{2} = 7 \Rightarrow m = \frac{95 - 4(7) - 3(15)}{2} = 11$

⇒ Solution is (7, 11, 15).

Let e = 17: $g = \frac{89 - 5(17)}{2} = 2 \Rightarrow m = \frac{95 - 4(2) - 3(17)}{2} = 18$

⇒ Solution is (2, 18, 17).

The solutions to this will always be negative if g or e is large.

Still negative, but much closer to 0.

Notice the pattern in the three solutions so far for (g, m, e):

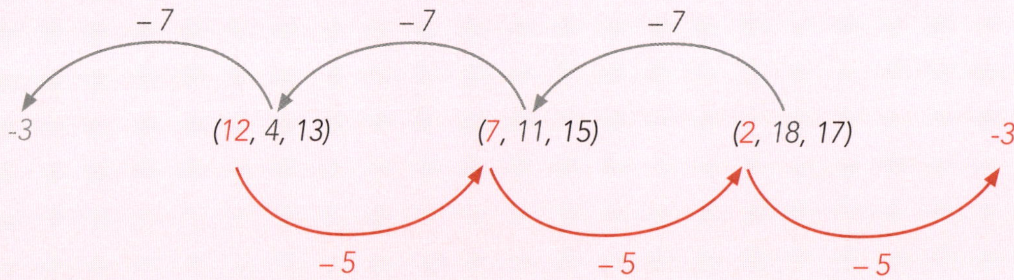

Extending the pattern of 'g's to the right or 'm's to the left produces negative numbers. Any solutions between these three will produce non-integral answers.
∴ These are the only three solutions to this system.

∴ They could make 12 batches of George's muffins, 4 of Maggie's and 13 of Eru's
or 7 batches of George's muffins, 11 of Maggie's and 15 of Eru's
or 2 batches of George's muffins, 18 of Maggie's and 17 of Eru's.

 ISBN: 9780170389426

Solve the following problems.

1 Three positive integers add to 18. The smallest is one fifth of the sum of the other two numbers. Let x represent the largest number, y represent the middle number and z represent the smallest.

a Write two equations to represent this situation.

b Let the third equation in this system be $x + y - 2z = k$. Investigate what happens when you attempt to solve these three equations when $k = 9$. Explain your result and describe the geometric arrangement made by the three planes. Find examples of solutions if they exist.

c Investigate what happens when you attempt to solve these three equations when $k \neq 9$. Explain your result and describe the geometric arrangement made by the three planes. Find examples of solutions if they exist.

These will not be asked for in future questions but you will still be expected to find them.

d Let the middle number be half of the difference between the largest and smallest number. Write a different third equation, and use all three equations to find the three numbers.

ISBN: 9780170389426

2 Three positive integers add to 36. If the middle number is added to the sum of triple the largest and four times the smallest, the total is 87. Let x represent the largest number, y represent the middle number and z represent the smallest.

a Write two equations to represent this situation.

b Let the third equation in this system be $kx + 3y + 6z = 159$. Find a value for k which results in many solutions to this system and find three of these solutions. Explain your reasoning and describe the geometric arrangement made by the three planes.

c Investigate what happens when you attempt to solve these three equations when k is numbers other than the answer you found in **b**. Explain your result and describe the geometric arrangement made by the three planes.

  ISBN: 9780170389426

3 Four friends visited a department store where socks, T-shirts and pyjamas were on sale. The prices for all these items were in whole dollars only.

Aroha bought three pairs of socks, two T-shirts and a pair of pyjamas for $92.
Bertie bought six pairs of socks, five T-shirts and two pairs of pyjamas for $204.
Cassie bought one pair of socks, three T-shirts and three pairs of pyjamas for $134.
Doug bought four pairs of socks, one T-shirt and two pairs of pyjamas for $106.

a One of them used a discount voucher. Who was it and what was it worth?

b Three friends visited a different department store where socks, T-shirts and pyjamas were on sale. One of these friends had a discount voucher for socks. Who was it, and what was it worth? Ellie bought five pairs of socks, two T-shirts and three pairs of pyjamas, which cost $160. Frank bought two pairs of socks, one T-shirt and one pair of pyjamas, which cost $58. Geoff bought three pairs of socks, one T-shirt and two pairs of pyjamas, which cost $98.

c Write a set of equations for this situation, assuming that the friend did **not** have a discount voucher. Explain why the system cannot be solved to give a unique solution, and describe the relationship between the three planes that represent these equations. Find some possible prices for socks, T-shirts and pyjamas.

ISBN: 9780170389426

4 Moana is making tiny bunches of flowers to go in the boys' buttonholes for the school formal. She has 25 rosebuds, 13 fern leaves and 19 lavender sprigs. She makes three different designs, and wants to use up all her flowers:

Design A needs three rosebuds, one fern leaf and two lavender sprigs.
Design B needs one rosebud, one fern leaf and one lavender sprig.
Design C needs four rosebuds, two fern leaves and three lavender sprigs.

a Write three equations to represent this situation.

b She tried to solve this system of equations on her calculator, but she got 'Ma ERROR'. Explain fully why this occurred, and describe the relationship between the three planes.

c Find possible solutions to these equations, and use these to write down at least three combinations of each type of design that could be made with the flowers available.

d Her friend suggests that she just puts two rosebuds in design A. Investigate how this will change the number of each design she makes.

e What would happen if she put four, five or six rosebuds in design A? Explain why this occurs.

ISBN: 9780170389426

5 Daniel assembles electronic kits, and his stocks of diodes, resistors and switches are running low. He has 127 diodes, 117 resistors and 69 switches left. There are plenty of all other components.

Kit A requires seven diodes, six resistors and four switches.
Kit B requires two diodes, four resistors and one switch.
Kit C requires six diodes, five resistors and three switches.

a He would like to assemble as many kits as possible, and use up all the diodes, resistors and switches. Write a set of equations to represent this situation, and solve it to determine the number of each type of kit that he should assemble.

b He checks the specifications for each kit, and finds he has made a mistake. Kit A contains eight diodes, not seven. Investigate how this change affects his situation.

c He rechecks his stocks, and finds a box containing an additional 11 diodes. Assuming that kit A does need eight diodes, not seven, investigate fully how this change affects his situation.

Practice tasks

Practice task one

Bags of lollies

The school council is going to sell bags of lollies for Gala Day. They have had 95 lollipops, 98 pineapple lumps and 278 jellybeans donated to them. They want to use all of these to make big, medium and small bags of lollies.

- Big bags contain four lollipops, five pineapple lumps and 11 jellybeans.
- Medium bags contain three lollipops, two pineapple lumps and nine jellybeans.
- Small bags contain two lollipops, three pineapple lumps and six jellybeans.

This activity requires you to apply systems of simultaneous equations to investigate the numbers of each size of bag that they can make.

Using the constraints outlined below, write a report that includes the information needed by the council.

Your report needs to answer the following questions:

- How many of each size of bag they should make?
- The council chairperson decides that the big bags don't have enough jellybeans, so she decides to put 12 jellybeans, rather than 11, in each big bag. Investigate how this change affects the situation. Fully justify your findings to the council.
- They recount the jellybeans and discover that they actually have 285, not 278. Assuming that they put 12 jellybeans in the big bags, investigate how this change affects the situation.

As you write your report, take care to clearly communicate your findings, using appropriate mathematical statements. Include relevant equations and calculations.

ISBN: 9780170389426

Practice task two

Phone plans
Liam is investigating phone plans from different companies. He wants to work out how much each plan charges for data, calling time and texts so that he can decide which to buy. He is only interested in costs for calling time per minute and texts to the nearest half cent.

He assumes that each company charges the same amount for each GB of data, each minute of calling time and each text, regardless of which plan he selects.

Smartaphone has three plans:
Plan A costs $27 per month and gives 1 GB of data, 200 minutes talking and 200 texts.
Plan B costs $37.50 per month and gives 1.5 GB of data, 200 minutes talking and 400 texts.
Plan C costs $61.50 per month and gives 2.5 GB of data, 400 minutes talking and 500 texts.

Catapult also has three plans:
Plan P costs $29 per month and gives 1 GB of data, 150 minutes talking and 300 texts.
Plan Q costs $39 per month and gives 1.5 GB of data, 250 minutes talking and 300 texts.
Plan R costs $69 per month and gives 2.5 GB of data, 400 minutes talking and 600 texts.

This activity requires you to apply systems of simultaneous equations to investigate the prices of data, calling time and texts under the various plans.

Using the constraints outlined below, write a report that includes the information that Liam wants.

Your report needs to answer the following:

- How much does Smartaphone charge for each GB of data, each minute of calling time and each text?
- Investigate the prices that Catapult charges for each minute of calling time and each text. Fully justify your findings to Liam.
- Catapult reduces the price for Plan R to $68. Investigate how this change affects the situation.

As you write your report, take care to clearly communicate your findings, using appropriate mathematical statements. Include relevant equations and calculations.

ISBN: 9780170389426

Practice task three

Electricity used by appliances

Tui lives in a flat. She is concerned that the power bill is high, and would like to know how much electricity is used by some appliances, particularly the oven, the TV and the heat pump. She can read their meter to determine total power use (in kilowatt hours), so she carries out some trials. She turns off all other appliances and records how long (hours) the oven, TV and heat pump are used for and the electricity used. Her results are shown in the table:

	Day 1	Day 2	Day 3
Oven	3	2	1
TV	2	3	1
Heat pump	5	8	2
Units used (kWh)	17.1	18.9	6.3

Her friend Cleo lives down the road and thinks this is a great idea, so she does the same for her appliances. Her results are shown in the table:

	Day 1	Day 2	Day 3
Oven	1	2	0.5
TV	2	4	1
Heat pump	5	9	2
Units used (kWh)	8.3	15.6	3.85

This activity requires you to apply systems of simultaneous equations to investigate the cost of running each appliance.

Using the constraints outlined below, write a report that answers the following:

- How much electricity does each of Tui's appliances use?
- Investigate the electricity used by each of Cleo's appliances. Explain fully why she might have difficulty determining this.
- She rechecked the piece of paper on which she had calculated the electricity totals and found that she had made a mistake. The total of 3.85 kWh on day three should have been 3.65 kWh. Investigate how this change affects the situation.

As you write your report, take care to clearly communicate your findings, using appropriate mathematical statements. Include relevant equations and calculations.

ISBN: 9780170389426

Answers

Solving 2 x 2 simultaneous equations (pp. 5–12)

1 Plotting (pp. 5–7)

1

x	2x + 1	y	Point
0	2(0) + 1	1	(0, 1)
1	2(1) + 1	3	(1, 3)
2	2(2) + 1	5	(2, 5)
3	2(3) + 1	7	(3, 7)
4	2(4) + 1	9	(4, 9)

x	-2x + 7	y	Point
0	-2(0) + 7	7	(0, 7)
1	-2(1) + 7	5	(1, 5)
2	-2(2) + 7	3	(2, 3)
3	-2(3) + 7	1	(3, 1)
4	-2(4) + 7	-1	(4, -1)

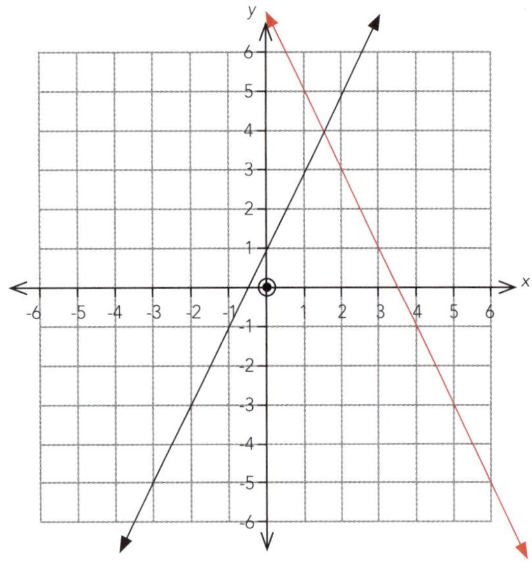

(1.5, 4)

2

x	3x – 1	y	Point
0	3(0) – 1	-1	(0, -1)
1	3(1) – 1	2	(1, 2)
2	3(2) – 1	5	(2, 5)
3	3(3) – 1	8	(3, 8)
4	3(4) – 1	11	(4, 11)

x	0.5x +4	y	Point
0	0.5(0) + 4	4	(0, 4)
1	0.5(1) + 4	4.5	(1, 4.5)
2	0.5(2) + 4	5	(2, 5)
3	0.5(3) + 4	5.5	(3, 5.5)
4	0.5(4) + 4	6	(4, 6)

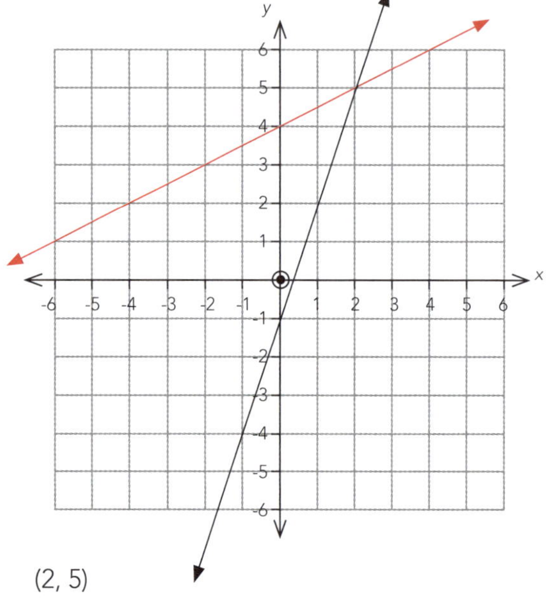

(2, 5)

2 Calculation (pp. 8–12)

a Substitution (pp. 8–9)

1	(6, 5)	**2**	(5, 11)
3	(21, 4)	**4**	(4, 14)
5	(-4, 3)	**6**	(-4, 9)
7	(-1, 2)	**8**	(-9, -2)
9	(-1, -2)	**10**	(-7.2, -2.3)

b Elimination (pp. 10–11)

1	(4, 10)	**2**	(-1, 6)
3	(3.75, -0.5)	**4**	(0, 7.5)
5	(-9, 8)	**6**	(2.5, 0.75)
7	(1, 2)	**8**	(14.8, -4.2)
9	(-3, 6)	**10**	(-102, 49)

Mixing it up (p. 12)

1	(31, -3)	**2**	(-5, 8)
3	(6.6, 3.8)	**4**	(2.4, -0.8)
5	(3.2, 4.6)	**6**	(5.5, -4.5)
7	(0.2, 1.6)	**8**	(-6, -12)
9	(0, 2)	**10**	(-2, 4)

ISBN: 9780170389426

Solving 3 x 3 simultaneous equations (pp. 13–18)

1 Substitution (pp. 13–14)

1 (1, -2, -3) **2** (0, 4, 0)
3 (5, 8, -3) **4** (26, -4, 14)

2 Elimination (pp. 15–16)

1 (-2, -5, 4) **2** (8, -1, 0)
3 (-1.75, 2.5, 0.25) **4** (11, 15, 4)

Mixing it up (pp. 17–18)

1 (1, 2.5, -3) **2** (33, -83, 51)
3 (3, 1, 3) **4** (-20, 40, -55)
5 (110, -27, 41) **6** (2, -6, 6)
7 (11.2, 5.6, 8.8) **8** (-17, -19, 14)

Using your calculator (pp. 19–22)

1 Solving 2 x 2 simultaneous equations (pp. 19–20)

1 (-2, 5) **2** (67, 48)
3 (17, -11) **4** (15, -12)
5 (-0.3, 1.6) **6** (-1, 0)
7 (-1, 1.5) **8** (-2, -5)
9 (1, -0.2) **10** (-1, 0)

2 Solving 3 x 3 simultaneous equations (pp. 21–22)

1 (-2, 15, 14) **2** (2, 7, 1)
3 (-4, 3, 2) **4** (5, 1, 3)
5 (2, 0, 1) **6** (5, 0, 1)
7 (5, 1, -2) **8** (1, 0, -3)
9 (2, 0, -3) **10** (3, 6, 0)
11 (15, 21, 34) **12** (-4, 2, 1)

Equations without a unique solution (pp. 23–37)

2D equations where there is not a unique solution (pp. 23–26)

1 Dependent: ① x 2 = ②
2 Real solution: (-1, 6)
3 Real solution: (2, 3)
4 Inconsistent
5 Dependent: ① = ② x 2
6 Dependent: ① x -3 = ②
7 Real solution: (-36, -16)
8 Inconsistent
9 Inconsistent
10 Real solution: (2, 4)
11 Inconsistent
12 Dependent: ① x -2 = ②
13 Inconsistent
14 Real solution: (-2, -2)

15 $k = 4$ **16** $k = -12$

3D equations where there is not a unique solution (pp. 27–37)

1 Inconsistent equations (pp. 27–31)

1 ① + ② ⇒ $3x - y + z = 26$

$\frac{3}{3} = \frac{-1}{-1} = \frac{1}{1} \neq \frac{26}{5}$ so ① + ② produces a plane parallel to ③.

∴

2 $\frac{1}{2} = \frac{2}{4} = \frac{3}{6} \neq \frac{4}{9}$ so ① is parallel to ③.

$\frac{1}{1} \neq \frac{2}{5} \neq \frac{3}{7} \neq \frac{4}{8}$ so ① is not parallel to ②.

∴

3 $\frac{4}{2} = \frac{2}{1} = \frac{6}{3} \neq \frac{3}{16}$ so ① is parallel to ②.

$\frac{4}{6} \neq \frac{2}{3} \neq \frac{6}{9} \neq \frac{3}{9}$ so ① is parallel to ③.

∴

4 ② + ③ ⇒ $3x + 2y + 4z = 11$

$\frac{3}{3} = \frac{2}{2} = \frac{4}{4} \neq \frac{11}{1}$ so ② + ③ produces a plane parallel to ①.

∴

2 Dependent equations (pp. 32–37)

1 2 x ① = ③ so these two planes are the same.

$\frac{1}{1} \neq \frac{2}{1} \neq \frac{3}{-2} \neq \frac{4}{3}$ so ① is not parallel to ②.

∴

2 Equations are linear combinations of each other: ① + ③ = ②.

∴

ISBN: 9780170389426

3 Equations are linear combinations of each other: 2 x ① + ② = ③.

∴

4 Equations are linear combinations of each other: 4 x ① – ② = ③.

∴

5 ② = 2 x ① and ③ = 4 x ① so all three planes are the same.

∴

6 Equations are linear combinations of each other: 3 x ① + 2 x ② = ③.

∴

7 Equations are linear combinations of each other: -5 x ① + -3 x ② = ③.

∴

8 2 x ① = ③ so these two planes are the same.

$\frac{2}{3} \neq \frac{1}{2} \neq \frac{-2}{5} \neq \frac{5}{5}$ so ① is not parallel to ②.

∴

Finding multiple solutions for dependent equations (pp. 38–41)

1 2 x ① = ② so these two planes are the same.
① is not parallel to ③.
(0, 5, 10), (1, 6, 9), (2, 7, 8), etc.

2 Equations are linear combinations of each other: 4 x ① – ② = ③.
(0, 10, 3), (1, 9, 4), (2, 8, 5), etc.

3 Equations are linear combinations of each other: 2 x ① + ② = ③.
(0, 8, 7), (2, 7, 8), (4, 6, 9), etc.

4 Equations are linear combinations of each other: ① + ③ = ②.
(0, 0, 10), (3, 1, 8), (6, 2, 6), (9, 3, 4), etc.

5 3 x ① = ③ so these two planes are the same.
① is not parallel to ②.
Only possibilities: (1, 11, 4), (1, 8, 5), (2, 5, 6), (3, 2, 7).

6 2 x ① = ③ so these two planes are the same.
① is not parallel to ②.
(0, 31, 1), (1, 27, 2), (2, 23, 3), etc.

7 Equations are linear combinations of each other: ① + ② = ③.
Only possibilities: (5, 1, 6), (7, 5, 1).

8 Equations are linear combinations of each other: ① + ③ = ②.
(10, 4, 3), (11, 21, 29), (12, 38, 55), etc.

Putting it together (pp. 43–45)

1 $\frac{2}{4} = \frac{-2}{-4} = \frac{6}{12} \neq \frac{10}{18}$ so ② is parallel to ③.

$\frac{3}{2} \neq \frac{-5}{-2} \neq \frac{12}{6} \neq \frac{-6}{10}$ so ① is not parallel to ②.

∴

inconsistent so no solutions.

2 Unique solution: (-4, 3, 2).

3 Equations are linear combinations of each other: ① x 5 – ② = ③.

∴

∴ dependent so many solutions.

4 ① + ② ⇒ $-3x - y - 3z = 37$

$\frac{-3}{3} = \frac{-1}{1} = \frac{-3}{3} \neq \frac{37}{19}$ so ① + ② produces a plane parallel to ③.

∴

∴ inconsistent so no solutions.

5 $\frac{-3}{4} = \frac{6}{-8} = \frac{-9}{12} \neq \frac{-1}{3}$ so ① is parallel to ③.

$\frac{-3}{3} \neq \frac{6}{-4} \neq \frac{-9}{6} \neq \frac{-1}{12}$ so ① is not parallel to ②.

∴

∴ inconsistent so no solutions.

 ISBN: 9780170389426

6 ③ x 1.25 = ① so these two planes are the same.

$\frac{20}{32} \neq \frac{10}{16} \neq \frac{35}{30} \neq \frac{255}{266}$ so ① is not parallel to ②.

∴ 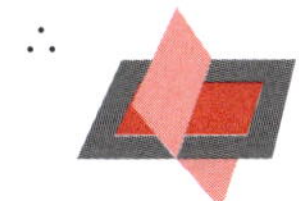

∴ dependent so many solutions.

7 ① – 3 x ② ⇒ $-4x + 2y - 3z = -32$

$\frac{-4}{-4} = \frac{2}{2} = \frac{-3}{-3} \neq \frac{-32}{8}$ so ① – 3 x ② produces a plane parallel to ③.

∴ 

∴ inconsistent so no solutions.

8 $\frac{3}{3} = \frac{-3}{-2} = \frac{-6}{-4} \neq \frac{-3}{10}$ so ① is parallel to ②.

$\frac{3}{-2} \neq \frac{-3}{3} \neq \frac{-6}{1} \neq \frac{-3}{7}$ so ① is not parallel to ③.

∴ 

∴ inconsistent so no solutions.

9 Equations are linear combinations of each other: ① + ② = 5 x ③.

∴ 

∴ dependent so many solutions.

10 Unique solution: (84, -7, -60)

11 ① + ② ⇒ $2x - 2y = -5$

$\frac{2}{2} = \frac{-2}{-2} \neq \frac{-5}{4}$ so ① + ② produces a line parallel to ③.

∴ 

∴ inconsistent so no solutions.

12 ① x 2 = ③, so these two planes are the same.

$\frac{2}{3} \neq \frac{1}{2} \neq \frac{-2}{5} \neq \frac{5}{5}$ so ① is not parallel to ②.

∴ 

∴ dependent so many solutions.

Forming and solving simultaneous equations (pp. 46–52)

1 Forming and solving 2 x 2 simultaneous equations (pp. 46–48)

1 Two numbers are 6 and 8.
2 Chairs cost \$120 and tables cost \$170.
3 183 adults' tickets and 334 children's tickets.
4 Julia has \$22 and Angus has \$13.
5 Ice creams cost \$3.20.
6 Kora is 37 and Jo is 11.
7 Apples cost \$1.25 per kg and carrots cost \$2.75 per kg.
8 Two numbers are 7 and 3.
9 Scored 22 two-point baskets and 12 three-point baskets.
10 Tops cost \$27 and pants cost \$35.
11 Sarah should pay \$25.50, Charlie should pay \$19.50.
12 There were seven spiders.

2 Forming and solving 3 x 3 simultaneous equations (pp. 49–52)

1 Apples cost \$3.50 per kg, broccoli costs \$1.50 per kg and carrots cost \$1.75 per kg.
2 She sells three small bags, five medium bags and seven large bags.
3 Basketballs cost \$42.99, cones cost \$2.99 and nets cost \$8.99.
4 Anna has \$220, Betty has \$160 and Carlos has \$190.
5 On Monday he got 17 messages, on Tuesday he got 69 and on Wednesday he got 23.
6 A pair of shorts cost \$15, singlets cost \$19 and the tracksuit cost \$78.
7 Kristen won 21 games, Erin won 18 and Hugh won 15.
8 Archer is 6, Xavier is 60 and Mason is 22.
9 Largest = 84°, middle-sized = 78° and smallest = 18°.
10 627
11 $y = 2x^2 - 3x + 7$
12 $y = x^3 + 4x^2 - 2x + 5$

Applications involving inconsistent and dependent equations (pp. 54–61)

1 a $x + y + z = 18$ ①
$x + y - 5z = 0$ ②

b $x + y - 2z = 9$ ③
①–②: $6z = 18 \Rightarrow 3z = 9 \Rightarrow z = 3$
①–③: $\underline{3z = 9}$
$0 = 0$
which is true ⇒ Equations are dependent so have multiple solutions.
Substituting $z = 3$ into ① ⇒ $x + y = 15$.
∴ Some solutions for this system are (0, 15, 3), (2, 13, 3), (3, 12, 3), …, (8, 7, 3), (9, 6, 3), (10, 5, 3), (11, 4, 3), (12, 3, 3), (13, 2, 3), (14, 1, 3), (15, 0, 3), etc.
However, only the red solutions are valid because *x* had to be biggest, *y* the middle value and *z* the smallest.

① + ② ⇒ $2x + 2y - 4z = 18$, which is the same as ③ ⇒ Planes ①, ② and ③ meet along the line $x + y = 15$.

Geometrically: 

c $x + y - 2z = 10$ ④
①–②: $6z = 18 \Rightarrow 3z = 9$
①–③: $\underline{3z = 8}$
$0 = 1$, which is false ⇒
Equations are inconsistent so have no solutions.

① + ② ⇒ $2x + 2y - 4z = 18$
⇒ $x + y - 2z = 9$. This plane is parallel to plane ④, but it is not the same plane. I know this because the coefficients are in proportion $\left(\frac{1}{1} = \frac{1}{1} = \frac{-2}{-2} \neq \frac{9}{8}\right)$, but the constants are not.

Geometrically: 

d The three numbers are 11, 4 and 3.

2 a $x + y + z = 36$ ①
$3x + y + 4z = 87$ ②

b $x + y + z = 36$ ①
$3x + y + 4z = 87$ ②
$kx + 3y + 6z = 159$ ③

None of the lines can be the same as each other because no equations have coefficients in proportion to each other.
∴ To have multiple solutions, equations must be linear combinations of each other.
③ = ② + 2 x ① ⇒ $k = 5$.

Solutions: (0, 19, 17), (3, 18, 15), (6, 17, 13), (9, 16, 11), (12, 15, 9), (15, 14, 7), (18, 13, 5), (21, 12, 3), (24, 11, 1).
However, only the red solutions are valid because *x* had to be biggest, *y* the middle value and *z* the smallest.

Geometrically: 

c For $k \neq 5$ the solution is always (0, 19, 17). Finding that the coefficient of *x* can be anything except 5 is not surprising because $x = 0$, which means that the size of its coefficient makes no difference to the total.
The three planes must meet at a point.

However, while (0, 19, 17) is a theoretical solution to the equations, it is not a real solution because *x* has to be the largest, which it clearly is not.

3 a $3s + 2t + p = 92$ ①
$6s + 5t + 2p = 204$ ②
$s + 3t + 3p = 134$ ③
$4s + t + 2p = 106$ ④

Attempting to solve any three of these involving ③ ⇒ equations which have solutions involving dollars and cents, rather than whole dollars.
∴ Cassie had the discount voucher.
Solving ①, ② and ④ ⇒ socks cost \$9, T-shirts cost \$20 and pyjamas cost \$25.

Substituting these into ③ ⇒ 9 + 60 + 75 = \$144 ∴ The voucher was for \$10.

b $5s + 2t + 3p = 160$ ①
$2s + t + p = 58$ ②
$3s + t + 2p = 98$ ③

① = ② + ③ for the coefficients, but not for the constants.
98 + 58 = 156, which is less than 160
⇒ person with the voucher was Frank or Geoff.

 ISBN: 9780170389426

② = ① – ③ for the coefficients, but not for the constants.
$160 - 98 = 62$, which is more than 58
⇒ person with the voucher was Frank.

He paid $4 less than the others for his socks, so the voucher was worth $4.

c If Frank had not had the voucher, he would have paid a total of $62.

$5s + 2t + 3p = 160$ ①
$2s + t + p = 62$ ②
$3s + t + 2p = 98$ ③

These equations are linear combinations of each other: ① = ② + ③.

∴ They look like: 

Possible prices:
(Socks $1, T-shirt $25, pyjamas $35)
(Socks $2, T-shirt $24, pyjamas $34)
(Socks $3, T-shirt $23, pyjamas $33)

4 **a** Rosebuds: $3a + b + 4c = 25$ ①
Fern leaves: $a + b + 2c = 13$ ② 
Lavender sprigs: $2a + b + 3c = 19$ ③

b Because ① + ② = 2 x ③. The equations are linear combinations of each other, so the equations are dependent — there are many solutions along the line where the three planes meet.

Geometrically: 

c Solutions: (6, 7, 0), (5, 6, 1), (4, 5, 2), (3, 4, 3), (2, 3, 4), (1, 2, 5), (0, 2, 6).
Note: These are the only solutions possible — any others would involve fractions of bunches or negative numbers.

d Rosebuds: $2a + b + 4c = 25$ ①
Fern leaves: $a + b + 2c = 13$ ②
Lavender sprigs: $2a + b + 3c = 19$ ③
There is one solution: (0, 1, 6).
However, this solution makes no sense, because she won't be making of Design A.

e Regardless of how many rosebuds she puts into Design A, you always get the same solution: (0, 1, 6). This occurs because this solution has her making none of Design A, so how many rosebuds she puts into it is irrelevant.

5 **a** Diodes: $7a + 2b + 6z = 127$ ①
Resistors: $6a + 4b + 5z = 117$ ②
Switches: $4a + b + 3z = 69$ ③
Solution:(11, 4, 7), so he should make 11 of Kit A, 4 of Kit B and 7 of Kit C.

b Diodes: $8a + 2b + 6z = 127$ ①
Resistors: $6a + 4b + 5z = 117$ ②
Switches: $4a + b + 3z = 69$ ③

From ① and ③: $\frac{8}{4} = \frac{2}{1} = \frac{6}{3} \neq \frac{127}{69}$

∴ ① and ③ are parallel but separate planes, but ② is not parallel.
∴ Equations are inconsistent and there are no solutions, so regardless of how many kits he makes, he cannot use up all these components.

Geometrically: 

c Diodes: $8a + 2b + 6z = 138$ ①
Resistors: $6a + 4b + 5z = 117$ ②
Switches: $4a + b + 3z = 69$ ③

From ① and ③: $\frac{8}{4} = \frac{2}{1} = \frac{6}{3} \neq \frac{138}{69}$

∴ ① and ③ are the same planes, but ② is a different plane which is not parallel.
∴ There will be infinite solutions along the line where ①, ② and ③ meet.

Geometrically: 

Only positive and integral solutions: (11, 4, 7) and (4, 2, 17).
Possible kits: 11 of A, 4 of B and 7 of C, or 4 of A, 2 of B and 17 of C.

Practice tasks (pp. 62–67)

Practice task one (pp. 62–63)

Bags of lollies

Lollipops: $4b + 3m + 2s = 95$ ①
Pineapple lumps: $5b + 2m + 3s = 98$ ②
Jellybeans: $11b + 9m + 6s = 278$ ③

Solution: (7, 15, 11), so they should make 7 big bags, 15 medium bags and 11 small bags.

Change to 12 jellybeans in big bags

Lollipops: $4b + 3m + 2s = 95$ ①
Pineapple lumps: $5b + 2m + 3s = 98$ ②
Jellybeans: $12b + 9m + 6s = 278$ ③

From ① and ③: $\frac{4}{12} = \frac{3}{9} = \frac{2}{6} \neq \frac{95}{273}$

∴ ① and ③ are parallel but separate planes, but ② is not parallel.

∴ Equations are inconsistent and there are no solutions, so regardless of how many of each size bag they make up, they cannot use up all the donated lollies.

Geometrically: 

Change to 285 jellybeans

Lollipops: $4b + 3m + 2s = 95$ ①
Pineapple lumps: $5b + 2m + 3s = 98$ ②
Jellybeans: $12b + 9m + 6s = 285$ ③

From ① and ③: $\frac{4}{12} = \frac{3}{9} = \frac{2}{6} = \frac{95}{285} = \frac{1}{3}$

∴ ① and ③ are the same planes, but ② is a different plane which is not parallel.

∴ There will be infinite solutions along the line where ①, ② and ③ meet.

Geometrically: 

To find some solutions

5 x ① – 4 x ② ⇒ $7m - 2s = 83$

$\therefore s = \frac{7m - 83}{2}$

From ①: $b = \frac{95 - 2s - 3m}{4}$

Let $m = 13$: $s = \frac{7(13) - 83}{2} = 4$

$\Rightarrow b = \frac{95 - 2(4) - 3(13)}{4} = 12$

⇒ **Solution is (12, 13, 4).**

Let $m = 15$: $s = \frac{7(15) - 83}{2} = 11$

$\Rightarrow b = \frac{95 - 2(11) - 3(15)}{4} = 7.$

⇒ **Solution is (7, 15, 11).**

Let $m = 17$: $s = \frac{7(17) - 83}{2} = 18$

$\Rightarrow b = \frac{95 - 2(18) - 3(17)}{4} = 2.$

⇒ **Solution is (2, 17, 18).**

Only positive and integral solutions for (b, m, s): (12, 13, 4), (7, 15, 11) and (2, 17, 18).
There cannot be any other solutions because following the pattern, if m is less than 13, s becomes negative. If m is bigger than 17, b becomes negative.
Possible numbers of bags:
12 big bags, 13 medium bags and 4 small bags, or 7 big bags, 15 medium bags and 11 small bags, or 2 big bags, 17 medium bags and 18 small bags.

Practice task two (pp. 64–65)
Phone plans

Let g represent the number of gigabytes of data.
Let m represent the number of minutes calling time.
Let t represent the number of texts.

Smartphone:

Plan A: $1g + 200m + 200t = 27$
Plan B: $1.5g + 200m + 400t = 37.5$
Plan C: $2.5g + 400m + 500t = 61.5$

Solution to this system: $g = 9$, $m = 0.06$, $t = 0.03$
∴ Under one of Smartphone's plans he has to pay $9 per GB of data, 6 cents per minute of calling time and 3 cents per text.

Catapult:

Plan P: $1g + 150m + 300t = 29$ ①
Plan Q: $1.5g + 250m + 300t = 39$ ②
Plan R: $2.5g + 400m + 600t = 69$ ③

This system of equations has no solution because the coefficients of the three equations are linear combinations of each other (① + ② = ③), but the constants are not: $29 + 39 \neq 69$. This means that the three planes represented by these equations never meet.

Geometrically: 

This means that in its three plans, Catapult does not charge the same amounts for each GB of data, each minute of calling time and each text, so it is not possible to give Liam a cost of each.

Catapult with reduced price for plan R:

Plan P: $1g + 300t + 150m = 29$ ①
Plan Q: $1.5g + 300t + 250m = 39$ ②
Plan R: $2.5g + 600t + 400m = 68$ ④

This system of equations has many solutions because the coefficients and the constants of the three equations are linear combinations of each other (① + ② = ④). This means that the three

 ISBN: 9780170389426

planes represented by these equations meet along a line where they all meet.

Geometrically: 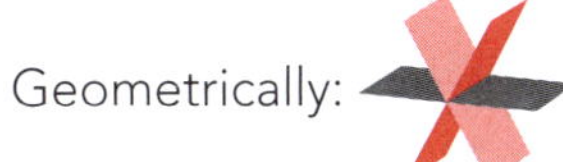

To find some solutions:

② – ①: $0.5g + 100m = 10$ ⑤

⑤ x 2: $g + 200m = 20$

$\therefore$ $\mathbf{g = 20 - 200m}$

From ①: $t = \dfrac{29 - 150m - g}{300}$

Let $m = 0.02$: $\Rightarrow g = 20 - 200(0.02) = 16$

$\Rightarrow t = \dfrac{29 - 150(0.02) - 16}{300} = 0.0\dot{3}$

Let $m = 0.03$: $\Rightarrow g = 20 - 200(0.03) = 14$

$\Rightarrow t = \dfrac{29 - 150(0.03) - 14}{300} = 0.035$

$\Rightarrow$ **Solution is (14, 0.035, 0.0$\dot{3}$).**

Let $m = 0.04$: $\Rightarrow g = 20 - 200(0.04) = 12$

$\Rightarrow t = \dfrac{29 - 150(0.04) - 12}{300} = 0.03\dot{6}$

Let $m = 0.05$: $\Rightarrow g = 20 - 200(0.05) = 10$

$\Rightarrow t = \dfrac{29 - 150(0.05) - 10}{300} = 0.038\dot{3}$

Let $m = 0.06$: $\Rightarrow g = 20 - 200(0.06) = 8$

$\Rightarrow t = \dfrac{29 - 150(0.06) - 8}{300} = 0.04$

$\Rightarrow$ **Solution is (8, 0.04, 0.06).**

Following the pattern in the value of t, I predict that the next value of m to produce a cost for t that is in half cents or more will be $m = 0.09$.

Let $m = 0.09$: $\Rightarrow g = 20 - 200(0.03) = 2$

$\Rightarrow m = \dfrac{29 - 150(0.09) - 2}{300} = 0.045$

$\Rightarrow$ **Solution is (2, 0.045, 0.09).**

Any further increase in m will produce $g = 0$ or less. Having a price of \$0 for data is not very realistic.

So the only solutions that are not negative or in less than half-cent values are:

1 (14, 0.035, 0.03), which means that data costs \$14 per GB, calling time costs 3 cents per minute and texts cost 3.5 cents each.
2 (8, 0.04, 0.06), which means that data costs \$8 per GB, calling time costs 6 cents per minute and texts cost 4 cents each.
3 (2, 0.045, 0.09), which means that data costs \$2 per GB, calling time costs 9 cents per minute and texts cost 4.5 cents each.

However, Liam would have no way of knowing which of these three options Catapult is using without further information.

Practice task three (pp. 66–67)
Electricity used by appliances

Let o represent the units of electricity used by the oven.
Let t represent the units of electricity used by the TV.
Let h represent the units of electricity used by the heat pump.

Tui:

Day 1: $3o + 2t + 5h = 17.1$
Day 2: $2o + 3t + 8h = 18.9$
Day 3: $1o + 1t + 2h = 6.3$

Solution to this system: $o = 3$, $t = 0.3$, $h = 1.5$

$\therefore$ The oven uses 3 kW per hour, the TV uses 0.3 kW per hour and the heat pump uses 1.5 kW per hour.

Cleo:

Day 1: $1o + 2t + 5h = 8.3$ ①
Day 2: $2o + 4t + 9h = 15.6$ ②
Day 3: $0.5o + 1t + 2h = 3.85$ ③

This system of equations has no real solution because the coefficients of the three equations are linear combinations of each other (① + 2 x ③ = ②), but the constants are not: $8.3 + 2 \times 3.85 \neq 15.6$. This means that the three planes represented by these equations never meet.

Geometrically: 

This means that Cleo cannot calculate the electricity use by each appliance because these equations have no solution. This might be because she didn't time their use accurately, she didn't calculate the total power use correctly, the power use of one or more appliance changed while she was measuring, or perhaps she ran some other appliances at the same time.

Cleo's power use with corrected value:

ISBN: 9780170389426

Day 1: $1o + 2t + 5h = 8.3$ ①
Day 2: $2o + 4t + 9h = 15.6$ ②
Day 3: $0.5o + 1t + 2h = 3.65$ ④

This system of equations has many solutions because the coefficients and the constants of the three equations are now linear combinations of each other (① + 2 x ④ = ②). This means that the three planes represented by these equations meet along a line where they all meet.

Geometrically: 

To find some solutions:
2 x ① – ②: $\boldsymbol{h = 1}$

From ④: $\boldsymbol{t = 3.65 - 2h - 0.5o}$
Let $h = 1$, $o = 1$: $\Rightarrow t = 3.65 - 2(1) - 0.5(1) = 1.15$
$\Rightarrow$ **Solution is (1, 1.15, 1).**

Let $h = 1$, $o = 2$: $\Rightarrow t = 3.65 - 2(1) - 0.5(2) = 0.65$
$\Rightarrow$ **Solution is (2, 0.65, 1).**

Let $h = 1$, $o = 3$: $\Rightarrow t = 3.65 - 2(1) - 0.5(3) = 0.15$
$\Rightarrow$ **Solution is (3, 0.15, 1).**
For all these solutions $h = 1$, which means that they all lie along the line $h = 1$. This means that the heat pump always used 1 kilowatt per hour.

1 (1, 1.15, 1), which means that the oven uses 1 kilowatt per hour, the TV uses 1.15 kilowatts per hour and the heat pump uses 1 kilowatt per hour.
2 (2, 0.65, 1), which means that the oven uses 2 kilowatts per hour, the TV uses 0.65 kilowatts per hour and the heat pump uses 1 kilowatt per hour.
3 (3, 0.15, 1), which means that the oven uses 3 kilowatts per hour, the TV uses 0.15 kilowatts per hour and the heat pump uses 1 kilowatt per hour.

Because the values do not have to be integers for electricity use, there will be many other solutions apart from these.

However, Cleo would have no way of knowing which of these combinations of electricity use is true without investigating further.

Pull-out section (centre of book)

Picture							
Identification	A	S	F	M	W	C	P
Result when you try to solve the system	V	K	K	K	B	B	B
Example	E	U	T	G	R	D	H
Solutions (if any)	N	X	X	X	L	Z	J

 ISBN: 9780170389426